The Contribution of Soil Science to the Development of and Implementation of Criteria and Indicators of Sustainable Forest Management

Related Society Publications

Carbon Forms and Functions in Forest Soils

Defining Soil Quality for a Sustainable Environment

Methods for Assessing Soil Quality

For information on these titles, please contact the ASA, CSSA, SSSA Headquarters Office; Attn.: Marketing; 677 South Segoe Road; Madison, WI 53711-1086. Telephone: (608) 273-8080, Fax: (608) 273-8089.

The Contribution of Soil Science to the Development of and Implementation of Criteria and Indicators of Sustainable Forest Management

Proceedings of a symposium sponsored by the S-7 and S-11 divisions of the Soil Science Society of America, the USDA Forest Service Northeastern Forest Experiment Station, and the Woods Hole Research Center. The symposium was held in St. Louis, MO, 31 October, 1995.

Organizing Committee
Mary Beth Adams, Kilaparti Ramakrishna,
and Eric A. Davidson, chair

Editorial Committee
Mary Beth Adams, Kilaparti Ramakrishna,
and Eric A. Davidson, chair

Editor-in-Chief SSSA
Jerry M. Bigham

Managing Editor
David M. Kral

Associate Editor
Marian K. Viney

SSSA Special Publication Number 53

Soil Science Society of America
Madison, Wisconsin

1998

Cover: Cover design by Patricia Scullion;
photograph provided by Eric A. Davidson

Cover Image: Dr. Manuel Maass of the Universidad Nacional Autónoma de México presented this satellite image of the border between Mexico and Guatemala at the SSSA symposium from which this book arose. Dr. Maass made the point that, while we often do not respect the integrity of forests, we usually do respect international borders. The intensive cultivation on the Mexico side (lightly shaded areas) stops abruptly at the international border, where the native forest of the region (dark shading) remained largely intact on the Guatemalan side in 1986. Similarly, while the management of forests and agricultural lands are the domain of sovereign states, international cooperation will be needed, as outlined in this book, to achieve common goals of identifying useful and effective criteria and indicators of soil health and forest health.

This image is a subset of a LANDSAT MSS satellite image from January 15, 1986. This subset is about 37 km (E–W) by 41 km (N–S). The San Pedro River is notable on the left, in the Mexican side. The original image is courtesy of NASA's Stennis Space Center and Conservation International. We thank Thomas Stone of the Woods Hole Research Center for enhancing the image for use here.

Soil Science Society of America, Inc.
677 South Segoe Road, Madison, Wisconsin 53711 USA

Library of Congress Catalog Card Number: 98-061724

Printed in the United States of America

CONTENTS

CONTENTS

FOREWORD

Expanding human population and resource consumption bring mounting pressures on the world's forests to produce fuel, timber, pulpwood, clean and abundant water, wildlife, recreation, and the many values of wilderness. Meeting such demands requires close attention to the elements supporting the health and function of forest ecosystems. Soil, in its interplay of physical, chemical, and biological properties and processes, is a singularly important element. Therefore, soil science is central to sustainable forest management.

Soil is a strong candidate for providing key criteria and indicators of sustainable forestry. Forest soil scientists of five generations in North America–and longer in Europe–have studied soils as natural bodies in landscapes and as components of systems varying from lands managed extensively as in the Pacific Northwest, to those managed with near-agricultural intensity, as in pine, hybrid poplar, and eucalyptus plantations throughout the world. Cumulative findings from this body of work may provide a basis for defining criteria and indicators useful for evaluating forest soils. Authors of chapters in this volume bring broad backgrounds and understandings of basic soil properties and processes that sustain forest productivity. These chapters represent the contributions of many of the world's leading forest soil researchers to the ongoing quest of all members of the Soil Science Society of America to develop sustainable systems of land uses and management.

Gary W. Petersen
SSSA President

PREFACE

As we constantly refine our notions of best management practices in forestry, strive to conserve forest ecosystem functions, and search for definitions of sustainable development of forest resources, the status of forest soils and the services that they provide must be rigorously evaluated. Certain attributes of forest soils, such as good infiltration, ample soil organic matter, lack of compaction, and occurrence of faunal activity, are well recognized as good general indicators of a healthy soil resource. But how much of a good thing is enough? Can we move beyond these qualitative indicators of soil health or degradation, and can we develop quantitative indicators of well defined criteria of the status of forest soils throughout the world under the variety of circumstances of climate, vegetation, and land use history?

Several international diplomatic initiatives are underway to define criteria and indicators (C&I) of forests, including soils, that are likely to form the basis of international agreements on the management, conservation, and sustainable development of forests. If science is to lead policy, the time is ripe for forest soil scientists to examine the proposed C&I and to establish a sound scientific basis for their selection and implementation.

To this end, the Soil Science Society of America sponsored a symposium at its annual meetings in St. Louis, MO, in October 1995, where a group of forest soil scientists examined proposed C&I to offer their opinions as to whether the policy negotiations were on the right track and to offer suggestions for improvements where appropriate. These chapters have since been expanded and additional authors have contributed, resulting in the critical analysis presented in this volume.

The chapters in this volume sample only a modest fraction of the diversity of soils, environmental conditions, and socio-economic situations in the world's forests to which the proposed C&I processes will be applied. Several approaches are critically analyzed for areas as diverse as Indonesia, Russia, western Europe, Canada, and the southeastern and western USA. A consensus emerges that considerable scientific basis exists upon which the C&I approach might be based, although the development and effective implementation of C&I will require considerable investment of resources. Much can be learned from the effort, as measuring changes in soil properties has the potential to tell us a great deal about long-term sustainability of forest management practices and uses. There is always need for more research, and establishment of effective C&I of forest soils will certainly require much more effort on the part of researchers and managers. The objective of this volume is to lay out the scientific basis as it now stands, so that the expectations of policies based on implementation of C&I will be realistic in the short term, and so that research needs for further improvement can be identified for the long term.

Eric A. Davidson, editor
The Woods Hole Research Center

PREFACE

CONTRIBUTORS

Mary Beth Adams	Project Leader and Soil Scientist, USDA Forest Service, Northeastern Research Station, P O. Box 404, Parsons, WV 26287
James R. Boyle	Professor of Forestry and Soil Ecology, College of Forestry, Oregon State University, Corvallis, OR, 97331-5703
James A. Burger	Professor of Forest Soil Science, Department of Forestry, 228 Cheatham Hall, Virginia Polytechnic Institute and State University, Blacksburg, VA, 24061-0324
Eric A. Davidson	Associate Scientist, The Woods Hole Research Center, P.O. Box 296, Woods Hole, MA, 02543
Pieter H.B. De Visser	Ecologist, Department of Animal Sciences, Agricultural University, P.O. Box 338, 6700 AH Wageningen, the Netherlands
Wim De Vries	Soil Chemist, SC-DLO Winand Staring Centre, Marijkeweg 11, P.O. Box 125, 6700 AC Wageningen, the Netherlands
Vladimir N. Gorbachev	Professor of Soil Science, Krasnoyarsk State Agricultural University, P.O. Box 8750, Axademgorodok, Krasnoyarsk, 660036, Russia
Kurniatun Hairiah	Lecturer in the Soil Science Department, University of Brawijaya, Jl. Veteran, Malang 65145, Indonesia
Daniel L. Kelting	Graduate Research Assistant, Department of Forestry, 228 Cheatham Hall, Virginia Polytechnic Institute and State University, Blacksburg, VA, 24061-0324
Andrei P. Laletin	Professor of Biological Science, Institute of Forest, Siberian Branch of the Russian Academy of Sciences, P.O. Box 26779, Axademgoradok, Krasnoyarsk, 660036, Russia
Ian K. Morrison	Research Scientist, Canadian Forest Service, Great Lakes Forestry Centre, P.O. Box 490, Sault Ste Marie, ON, Canada P6A 5M7
Daniel Murdiyarso	Program Head, BIOTROP/GCTE Impacts Centre for Southeast Asia, Jl. Raya Tajur Km 6, P.O. Box 116, Bogor, Indonesia
Meine van Noordwijk	Principal Soil Ecologist, International Centre for Research in Agroforestry, P.O. Box 161, Bogor 16001, Indonesia
Robert F. Powers	Science Team Leader, USDA Forest Service, Pacific Southwest Research Station, 2400 Washington Avenue, Redding, CA 96001
R.J. Raison	Chief Research Scientist, CSIRO Forestry and Forest Products, P.O. Box E4008, Kingston, ACT 2604, Australia
Kilaparti Ramakrishna	Senior Associate, The Woods Hole Research Center, P.O. Box 296, Woods Hole, MA, 02543
C.T. Smith	Project Leader, New Zealand Forest Research Institute Ltd., Private Bag 3020, Rotorua, New Zealand
Allan E. Tiarks	Supervisory Soil Scientist, Southern Research Station, USDA Forest Service, 2500 Shreveport Highway, Pineville, LA, 71360
Nico van Breemen	Professor of Soil Formation and Ecopedology, Laboratory of Soil Science and Geology, Wageningen Agricultural University, P.O. Box 37, 6700 AA Wageningen, the Netherlands
Paul L. Woomer	Visiting Lecturer, Department of Soil Science, University of Nairobi, P.O. Box 30197, Nairobi, Kenya

Conversion Factors for SI and non-SI Units

Conversion Factors for SI and non-SI Units

To convert Column 1 into Column 2, multiply by	Column 1 SI Unit	Column 2 non-SI Units	To convert Column 2 into Column 1, multiply by
		Length	
0.621	kilometer, km (10^3 m)	mile, mi	1.609
1.094	meter, m	yard, yd	0.914
3.28	meter, m	foot, ft	0.304
1.0	micrometer, μm (10^{-6} m)	micron, μ	1.0
3.94×10^{-2}	millimeter, mm (10^{-3} m)	inch, in	25.4
10	nanometer, nm (10^{-9} m)	Angstrom, Å	0.1
		Area	
2.47	hectare, ha	acre	0.405
247	square kilometer, km^2 (10^3 m)^2	acre	4.05×10^{-3}
0.386	square kilometer, km^2 (10^3 m)^2	square mile, mi^2	2.590
2.47×10^{-4}	square meter, m^2	acre	4.05×10^3
10.76	square meter, m^2	square foot, ft^2	9.29×10^{-2}
1.55×10^{-3}	square millimeter, mm^2 (10^{-3} m)^2	square inch, in^2	645
		Volume	
9.73×10^{-3}	cubic meter, m^3	acre-inch	102.8
35.3	cubic meter, m^3	cubic foot, ft^3	2.83×10^{-2}
6.10×10^4	cubic meter, m^3	cubic inch, in^3	1.64×10^{-5}
2.84×10^{-2}	liter, L (10^{-3} m^3)	bushel, bu	35.24
1.057	liter, L (10^{-3} m^3)	quart (liquid), qt	0.946
3.53×10^{-2}	liter, L (10^{-3} m^3)	cubic foot, ft^3	28.3
0.265	liter, L (10^{-3} m^3)	gallon	3.78
33.78	liter, L (10^{-3} m^3)	ounce (fluid), oz	2.96×10^{-2}
2.11	liter, L (10^{-3} m^3)	pint (fluid), pt	0.473

		Mass	
2.20×10^{-3}	gram, g (10^{-3} kg)	pound, lb	454
3.52×10^{-2}	gram, g (10^{-3} kg)	ounce (avdp), oz	28.4
2.205	kilogram, kg	pound, lb	0.454
0.01	kilogram, kg	quintal (metric), q	100
1.10×10^{-3}	kilogram, kg	ton (2000 lb), ton	907
1.102	megagram, Mg (tonne)	ton (U.S.), ton	0.907
1.102	tonne, t	ton (U.S.), ton	0.907
		Yield and Rate	
0.893	kilogram per hectare, kg ha^{-1}	pound per acre, lb $acre^{-1}$	1.12
7.77×10^{-2}	kilogram per cubic meter, kg m^{-3}	pound per bushel, lb bu^{-1}	12.87
1.49×10^{-2}	kilogram per hectare, kg ha^{-1}	bushel per acre, 60 lb	67.19
1.59×10^{-2}	kilogram per hectare, kg ha^{-1}	bushel per acre, 56 lb	62.71
1.86×10^{-2}	kilogram per hectare, kg ha^{-1}	bushel per acre, 48 lb	53.75
0.107	liter per hectare, L ha^{-1}	gallon per acre	9.35
893	tonne per hectare, t ha^{-1}	pound per acre, lb $acre^{-1}$	1.12×10^{-3}
893	megagram per hectare, Mg ha^{-1}	pound per acre, lb $acre^{-1}$	1.12×10^{-3}
0.446	megagram per hectare, Mg ha^{-1}	ton (2000 lb) per acre, ton $acre^{-1}$	2.24
2.24	meter per second, m s^{-1}	mile per hour	0.447
		Specific Surface	
10	square meter per kilogram, m^2 kg^{-1}	square centimeter per gram, cm^2 g^{-1}	0.1
1000	square meter per kilogram, m^2 kg^{-1}	square millimeter per gram, mm^2 g^{-1}	0.001
		Pressure	
9.90	megapascal, MPa (10^6 Pa)	atmosphere	0.101
10	megapascal, MPa (10^6 Pa)	bar	0.1
1.00	megagram, per cubic meter, Mg m^{-3}	gram per cubic centimeter, g cm^{-3}	1.00
2.09×10^{-2}	pascal, Pa	pound per square foot, lb ft^{-2}	47.9
1.45×10^{-4}	pascal, Pa	pound per square inch, lb in^{-2}	6.90×10^3

(continued on next page)

Conversion Factors for SI and non-SI Units

To convert Column 1 into Column 2, multiply by	Column 1 SI Unit	Column 2 non-SI Units	To convert Column 2 into Column 1, multiply by
		Temperature	
1.00 (K – 273)	kelvin, K	Celsius, °C	1.00 (°C + 273)
(9/5 °C) + 32	Celsius, °C	Fahrenheit, °F	5/9 (°F – 32)
		Energy, Work, Quantity of Heat	
9.52×10^{-4}	joule, J	British thermal unit, Btu	1.05×10^{3}
0.239	joule, J	calorie, cal	4.19
10^{7}	joule, J	erg	10^{-7}
0.735	joule, J	foot-pound	1.36
2.387×10^{-5}	joule per square meter, J m^{-2}	calorie per square centimeter (langley)	4.19×10^{4}
10^{5}	newton, N	dyne	10^{-5}
1.43×10^{-3}	watt per square meter, W m^{-2}	calorie per square centimeter minute (irradiance), cal cm^{-2} min^{-1}	698
		Transpiration and Photosynthesis	
3.60×10^{-2}	milligram per square meter second, mg m^{-2} s^{-1}	gram per square decimeter hour, g dm^{-2} h^{-1}	27.8
5.56×10^{-3}	milligram (H_2O) per square meter second, mg m^{-2} s^{-1}	micromole (H_2O) per square centimeter second, µmol cm^{-2} s^{-1}	180
10^{-4}	milligram per square meter second, mg m^{-2} s^{-1}	milligram per square centimeter second, mg cm^{-2} s^{-1}	10^{4}
35.97	milligram per square meter second, mg m^{-2} s^{-1}	milligram per square decimeter hour, mg dm^{-2} h^{-1}	2.78×10^{-2}
		Plane Angle	
57.3	radian, rad	degrees (angle), °	1.75×10^{-2}

	Electrical Conductivity, Electricity, and Magnetism		
10	siemen per meter, S m^{-1}	millimho per centimeter, mmho cm^{-1}	0.1
10^4	tesla, T	gauss, G	10^{-4}
	Water Measurement		
9.73×10^{-3}	cubic meter, m^3	acre-inch, acre-in	102.8
9.81×10^{-3}	cubic meter per hour, m^3 h^{-1}	cubic foot per second, ft^3 s^{-1}	101.9
4.40	cubic meter per hour, m^3 h¯1	U.S. gallon per minute, gal min^{-1}	0.227
8.11	hectare meter, ha m	acre-foot, acre-ft	0.123
97.28	hectare meter, ha m	acre-inch, acre-in	1.03×10^{-2}
8.1×10^{-2}	hectare centimeter, ha cm	acre-foot, acre-ft	12.33
	Concentrations		
1	centimole per kilogram, cmol kg^{-1}	milliequivalent per 100 grams, meq 100 g^{-1}	1
0.1	gram per kilogram, g kg^{-1}	percent, %	10
1	milligram per kilogram, mg kg^{-1}	parts per million, ppm	1
	Radioactivity		
2.7×10^{-11}	becquerel, Bq	curie, Ci	3.7×10^{10}
2.7×10^{-2}	becquerel per kilogram, Bq kg^{-1}	picocurie per gram, pCi g^{-1}	37
100	gray, Gy (absorbed dose)	rad, rd	0.01
100	sievert, Sv (equivalent dose)	rem (roentgen equivalent man)	0.01
	Plant Nutrient Conversion		
	Elemental	***Oxide***	
2.29	P	P_2O_5	0.437
1.20	K	K_2O	0.830
1.39	Ca	CaO	0.715
1.66	Mg	MgO	0.602

1 Intergovernmental Negotiations on Criteria and Indicators for the Management, Conservation, and Sustainable Development of Forests: What Role for Soil Scientists?

Kilaparti Ramakrishna and Eric A. Davidson

Woods Hole Research Center
Woods Hole, Massachusetts

When the topic of conservation and better use of world forest resources burst open on the world arena during preparations for the 1992 United Nations Conference on Environment and Development (alias, the Earth Summit), the contours of the problem were not clearly defined. World forest resources had not previously been the subject of discussions at the United Nations. At a G-7 summit (Canada, France, Germany, Italy, Japan, England, and the USA) in 1990, the seven most industrialized countries proposed the adoption of a world forest agreement, which might have taken the form of a legally binding agreement at the Earth Summit. The developing countries opposed this proposal, however, because of their concerns about the timing and scope of an agreement, about safeguarding their national sovereignty, and about protection of timber trade and economic development (Ramakrishna, 1993). The developing countries believed that the industrialized countries were interested in pointing fingers at the developing countries for permitting tropical deforestation and in trying to set aside tropical forest land as preserves to offset CO_2 emissions by industrialized countries (Maini & Ullsten, 1993). The Earth Summit was therefore able to agree: (i) on a set of *non-legally binding* but *authoritative* principles on forest management and (ii) on the need to *combat deforestation* (Chapter 11 of Agenda 21; United Nations, 1992).

As part of the principles enumerated at the Earth Summit, governments agreed to pursue the "formulation of scientifically sound criteria and guidelines for the management, conservation and sustainable development of all types of forests." This agreement has resulted in several processes at national, regional, and interregional levels focused on developing criteria and indicators (C&I) for

---

The Contribution of Soil Science to the Development of and Implementation of Criteria and Indicators of Sustainable Forest Management. SSSA Special Publication no. 53.

the conservation and sustainable management of boreal, temperate and tropical forests. The leading C&I processes are reviewed in this chapter.

Because the soil resource is critical for maintaining healthy forests, the C&I pertaining to soils deserve particular attention in terms of their soundness of scientific basis, effectiveness, and feasibility. With the adoption of the Vienna Convention on Ozone Layer Protection and its Montreal Protocol and the Framework Convention on Climate Change and its Kyoto Protocol, there has been a new recognition of the critical importance of scientists in articulating problems in a policy context and incorporating them as an integral arm for the purposes of further strengthening conventions and protocols. To this end, the Soil Science Society of America sponsored a symposium at its annual meetings in St. Louis, MO, in October, 1995, where a group of forest soil scientists examined proposed C&I to offer their opinions as to whether the policy negotiations were on the right track and to offer suggestions for improvements where appropriate. These articles have since been expanded and additional authors have contributed, resulting in the critical analysis presented in this volume.

THE GLOBAL IMPORTANCE OF FORESTS

Economic Benefits

Both industrialized and developing countries have a clear enough sense of the role timber trade plays in their economy. There are several statistics to prove this point, but a particularly telling one is advanced by the Food and Agriculture Organization (FAO). According to this source, the annual contribution of forest products to the world economy reached some US$400 billion in 1991, or about 2% of the global economic product (FAO, 1994). In terms of the number of individuals employed, one estimate suggests that forestry and logging currently provide the equivalent of 60 million work-years worldwide, of which about 80% are in developing countries. Revenues from timber trade is often the principal way of ensuring that there is balance of payments in international trade (Repetto, 1993). While this is a recognition, in a traditional sense, of the economic value of forest resources, its environmental services of nontimber forest products, tourism, and others, were widely recognized only recently and slowly.

Environmental Services

Concern is increasing that the temperate and boreal forests of many northern developed countries are being degraded by a combination of industrial pollution, climate change, land use change, forest fires, disease, and other factors (Woodwell, 1993). Two independent surveys of tropical deforestation conducted by the National Research Council of the United States and the United Nations FAO/United Nations Environment Programme (UNEP) came to a similar conclusion in their assessments for the tropics (Houghton, 1993). According to these studies, the figures of deforestation in closed forests in 1980 amounted to 7.3 $\times 10^6$ ha yr^{-1}, and combining closed and open forests amounted in the same year to 11.3×10^6 ha yr^{-1}. This astonishing figure was subsequently revised to 15.4×10^6 ha yr^{-1} by FAO in 1993. This rate of deforestation occurred despite US$1 billion of national government and bilateral aid directed at the problem of sustainable forestry development (Oksanen et al., 1993). Only recently has the recognition been gaining ground that forests provide a wide range of environmental benefits and values in addition to the socioeconomic values. Shrinking of global forest cover and ecological deterioration of former forest lands already are having a negative impact on global climate, habitat for genetic diversity, soil and water resources, and on human welfare (United Nations, 1995a).

Because of their multiple roles, forests have emerged as a priority issue in international scientific, socioeconomic, political, environmental, and trade agendas. While forests are located within the boundaries of sovereign nations, many of their ecological benefits transcend national boundaries. The challenge is to reconcile nations' legitimate rights to develop and conserve their resources to meet their national policy objectives with their responsibility to maintain a healthy regional and global environment.

POLICY DEVELOPMENTS

Earth Summit

Despite compelling reasons regarding the economic and environmental importance of forests, forests have not been given the attention that they deserve by the United Nations. Early attempts largely focused on national efforts that eventually encompassed the tropical forested countries. The Earth Summit preparations clearly gave an opportunity to address forests truly as matters of global concern. The Earth Summit adopted Agenda 21 (United Nations, 1992), containing an action programme for the world community to the year 2000 and beyond. The Agenda, while highlighting issues that have traditionally engaged the developing countries (viz., the need for new and additional resources), also emphasized the importance of involving all stakeholders. As with most United Nations resolutions, it left the primary responsibility for follow-up to the nations. Amongst the 40 chapters, Chapter 11 is the central chapter of the Agenda 21 that pertains to forests—it covers the following program areas:

- sustaining the multiple roles and functions of all types of forests, forest lands and woodlands;
- enhancing the protection, sustainable management and conservation of forests, and the greening of degraded areas, through forest rehabilitation, afforestation, reforestation and other rehabilitative measures;
- promoting efficient use and assessment to recover the full valuation of the goods and services provided by forests, forest lands, and woodlands; and
- establishing and/or strengthening capacities for the planning, assessment, and systematic observations of forests and related programmes, projects, and activities including commercial trade and processes.

The most important document negotiated at the Earth Summit on forests was "the non-legally binding authoritative statement of principles for a global consensus on the management, conservation and sustainable development of all types of forests." Agreement on these principles was reached only after intensive and often divisive negotiations that spilled into the Summit Meeting of Heads of State and Government in 1992. Recognizing the industrialized countries' interests in this issue, the developing countries have made their participation in the eventual agreement contingent upon obtaining financial assistance, technology transfer, and debt relief. These issues have therefore taken center stage in the context of the eventual agreement.

In any event, the after-taste of concluding business on the principles in 1992 was such that most of the actors thought it would be appropriate not to revisit the implementation or operational aspects of these principles for the next 2 yr. Consequently, in the thematic plan of work adopted by the United Nations General Assembly, the Assembly scheduled the topic of forest conservation and related matters to be addressed in 1995. Some considered this as a *ban* on further discussions until 1995 and others considered it as a *cool-off* period.

In the meantime, the International Tropical Timber Agreement (ITTA), which is a commodity agreement negotiated under the auspices of the United Nations Conference on Trade and Development, was renegotiated and given a new lease on life without expanding its scope beyond tropical timber products, to a wider geographic sphere, or to a more global environmental sphere.

Criteria and Indicator Processes

Putting off discussions on a forest agreement until 1995 was not satisfactory for many. Those who were anxious to begin work found comfort in one of the forest principles adopted at the Earth Summit that stated:

> "Sustainable forest management and use should be carried out in accordance with national development policies and priorities and on the basis of environmentally sound national guidelines. In the formulation of such guidelines, account should be taken, as appropriate and if applicable, of relevant internationally-agreed methodologies and criteria."

Those desirous of promoting an international agreement on forests identified the reference to methodologies and criteria as the hook on which to base further developments. A window was thus created through which an international agreement could be allowed to establish a set of general criteria for sustainable development and conservation of forests and a common set of methodologies (indicators) that might be used to assess progress towards those criteria, but national sovereignties and national priorities would still be respected because implementation would be left to national policies and guidelines.

The idea that criteria and indicators would be useful to determine sustainability of forest management found widespread favor and was the subject of a vigorous intergovernmental activity between 1992 and 1997 under at least six separate intergovernmental exercises. As a result of this interest in C&I, one could make the case that there has been a lot of movement in addressing forest issues despite the perceived ban imposed on further elaboration of these items at the Earth Summit and immediately thereafter.

These intergovernmental activities resulted in several agreements during the past years. In May 1992, the International Tropical Timber Organization (ITTO) developed and adopted criteria and indicators for sustainable tropical forest management based on signals from the tropical timber markets. This acted as a precursor and in some instances as a model for several initiatives that followed.

The Helsinki Process is a pan-European political process that began with the First Ministerial Conference on the Protection of Forests in Europe in December, 1990, and culminated in June, 1994, with the adoption of their version of criteria and indicators. At a meeting in January, 1995, the 29 European signatories reviewed the European experiences in using the criteria and associated indicators and agreed to convene the Third Ministerial Conference in 1998 for further evaluation and refinement (for complete details see Summary Report of the Intergovernmental Seminar on Criteria and Indicators for Sustainable Forest Management, Helsinki, Finland, 19–22 Aug. 1996; http://www.mmm.fi/isci/final.htm).

The Montreal Process is a Canadian initiative that began in September, 1993, with the goal of developing a scientifically rigorous set of criteria and indicators that could be used to measure forest management. With support from the USA, Canada tried to bring the Helsinki and Montreal processes together, but gave up this attempt when several prominent countries (France, Germany, and England) that are part of the Helsinki Process expressed their preference to remain independent. After several informal meetings, the Montreal Process was formalized and called itself the Working Group on Criteria and Indicators for the Conservation and Sustainable Management of Temperate and Boreal Forests. Participation now includes 12 countries: Argentina, Australia, Canada, Chile, China, Japan, Republic of Korea, Mexico, New Zealand, Russian Federation, USA, and Uruguay. In Santiago, Chile, on 2–3 Feb. 1995, this group adopted its non-legally binding criteria and indicators and recommended that other countries also adopt the criteria and indicators with the provision that, as scientific understanding increases, the criteria and indicators may be revised.

In contrast to the above processes, the Tarapoto Proposal to develop criteria and indicators arose within the framework of the Amazon Cooperation Treaty (ACT), a regional intergovernmental agreement, with membership drawn from Bolivia, Brazil, Colombia, Ecuador, Guyana, Peru, Suriname, and Venezuela. The ACT, signed in 1978 to promote harmonious development in the Amazon Basin, adopted the Tarapoto Proposal with the aim to combine "environmental sustainability factors with the optimal economic use of the Amazon forests and overall social development" (United Nations, 1995b; See also V.R. Carazo, Analysis and Prospects of the Tarapoto Proposal: Criteria and Indicators for the Sustainability of the Amazonian Forest, XI World Forestry Congress, Antalya, Turkey, 13–22 Oct. 1997; http://www.fao.org/waicent/faoinfo/forestry/wforcong/publi/V6/T373E/2.HTM#TOP; and see also J. Cesar Centeno, Criteria and Indicators for the Sustainable Management of Amazon Forests, International Conference on Global Approaches to Sustainable Forest Management,

Certification, Criteria and Indicators, Prince George, Canada, September, 1997; http://csf.colorado.edu/lists/elan/nov97/0044.html).

The UNEP and the FAO initiated a process similar to the above in Africa during 1995 and 1996 and proposed a set of criteria and indicators to be used at the national level in the CILSS (Intergovernmental Committee on Drought Control in the Sahel, comprising Burkina Faso, Cape Verde, Gambia, Guinea Bissau, Mali, Mauritania, Niger, Senegal, and Chad), the SADC (Southern African Development Community, comprising Angola, Zambia, Swaziland, Lesotho, Zimbabwe, Mauritius, Botswana, Mozambique, Malawi, Namibia, South Africa, and Tanzania), and the IGADD (Intergovernmental Authority on Drought and Development, comprising Djibouti, Ethiopia, Kenya, Somalia, Sudan, and Uganda) regions. The results, endorsed by the African Wildlife and Forestry Commission, are to be further refined in the countries concerned. These were reviewed at a Colloquium organized by the African Timber Organization during 16–17 Apr. 1997 (See Bai-Mass Taal, Criteria and Indicators for Sustainable Forest Management in Dry-Zone Africa, XI World Forestry Congress, Antalya, Turkey, 13–22 Oct. 1997, http://www.fao.org/waicent/faoinfo/forestry/wforcong/publi/V6/T374E/2.HTM#TOP).

The Lepaterique Process, initiated under the auspices of the Regional Treaty for Management and Conservation of Natural Forests Ecosystems and the Development of Forest Plantations signed in 1993 by the seven Central American countries (Belize, Guatemala, El Salvador, Honduras, Nicaragua, Costa Rica, and Panama), formulated the Central American approach to criteria and indicators for sustainable forest management. Along the lines of other processes, this too went through a series of regional workshops before forwarding its conclusions for adoption to the Summit of Presidents of the Central American countries in April, 1997 (See Juan Blas Zapata, The Central American Process of Sustainable Development, XI World Forestry Congress, Antalya, Turkey, 13–22 Oct. 1997, http://www.fao.org/waicent/faoinfo/forestry/wforcong/publi/V6/T375E/1.HTM#TOP).

The C&I are meant to be implemented at national levels, thus respecting national sovereignty and national needs and circumstances, while an international agreement on C&I would provide *harmony* among nations in terms of goals and guiding principles. This approach has resulted in a positive change in perspectives among the industrialized and developing countries. The resistance to beginning negotiations on a world forest agreement is less intense now than before. Many governmental representatives in intergovernmental discussions refer to C&I as providing the basis for an international agreement on forests.

Future Policy Steps

The United Nations Commission on Sustainable Development created an Intergovernmental Panel on Forests (IPF), which produced its report to the Fifth Session of the UN Commission on Sustainable Development in 1997 (United Nations, 1997a). Many countries had hoped that, once the IPF concluded its work in 1997, the left over business from the Earth Summit (viz., an international agreement on forests), would be taken up in earnest. Despite the fact that most of the countries that opposed the forest agreement in 1992 have either expressed their willingness to join negotiations for a world forest agreement, or were not going to put up any hurdles in its path, the IPF could not agree on a recommendation to the Commission on Sustainable Development in favor of an agreement. In the end, the IPF proposed the following three options: (i) to continue the intergovernmental policy dialogue; (ii) to establish an open-ended intergovernmental forum on forests; and (iii) to establish an intergovernmental negotiating committee on a legally binding instrument on all types of forest, with a focused and time-limited mandate. Despite progress on defining the issues, the IPF has been judged by many as largely a failure, due to its inability to make a recommendation on the further development of an international forest convention or other legal instrument (M. Holdgate and M. Rodriguez Becerra, co-chairs of the UNIPF, 1997, personal communication). In June, 1997, after reviewing the progress made since the Rio Summit in 1992, the United Nations General Assembly agreed to establish an Intergovernmental Forum on Forests (IFF) as a follow-up to the IPF, with a mandate that the IFF complete its work by the year 2000 (United Nations, 1997b).

As the global debate on world forests moves to a new phase, three initiatives stand out as crucial. These include, in addition to the IFF mentioned above, the WCFSD and ITFF. The InterAction Council of Former Heads of State and Government established an independent World Commission on Forests and Sustainable Development (WCFSD) in June, 1994. The Commission, working as a nongovernmental entity, will be completing its report in 1998 (http://iisd1.iisd.ca/wcfsd/). Many of the general areas of concern identified by the national and international community were looked at by the WCFSD through the so-called *Public Hearings*. Not only governmental representatives, but also all of the actors with any interest in how forest resources are managed, could share their views in these hearings and shape the final conclusions of the WCFSD. As the United Nations carried out its work through IPF, it became clear that some sort of donor agency coordination is necessary. The result was the establishment of an Interagency Task Force on Forests (ITFF) with representation from the Center for International Forestry Research, Food and Agriculture Organization, International Tropical Timber Organization, Secretariat of the Convention on Biological Diversity, United Nations Department for Policy Coordination and Sustainable Development, United Nations Development Programme, United Nations Environment Programme and the World Bank (gopher://gopher.un.org:70/00/esc/cn17/ipf/itffplan.txt).

These three bodies (IFF, WCFSD, and ITFF) will ensure that conservation and better utilization of world forest resources will engage the attention of the international policy community during their existence in the post-1997 era. These fora will continue to be in existence for some time, but the original emphasis placed on an international agreement will be delayed. This delay offers an opportunity for scientists to comment on the proposals being discussed, especially the much proliferated sets of criteria and indicators. These efforts could form the basis for a forest agreement and, more importantly, for attaining the elusive goal of sustainable forest management. If the scientific community concludes that the C&I processes are on the right track, saying so now will further strengthen the efforts already underway and could help bring the matter to a resolution as an agreement in a fairly short order. On the other hand, if scientists find serious inadequacies of C&I in achieving the hallowed notion of sustainability, a caution statement will ensure that a fresh look is taken on the whole issue.

One goal is to ensure that such concerns are presented to the above three fora, and particularly to the ITFF. The ITFF has developed an ambitious plan of action to the year 2000 and beyond. The first stage is to conclude in the year 2000 and the ITFF hopes to bring together countries engaged in all of the on-going processes on C&I. They hope to reach an agreement on the concepts and arrive at an overall compatibility among these processes.

EXISTING EVALUATIONS OF CRITERIA AND INDICATORS

The Center for International Forestry Research (CIFOR) undertook the first attempt to test the C&I at the field level (forest management unit level) in a variety of locations. The approach involved multidisciplinary teams of foresters, social scientists, and ecologists, selecting and evaluating C&I in an interdisciplinary fashion, in four locations around the world (Prabhu et al., 1996). The sites selected were Germany, Indonesia, Cote d'Ivoire, and Brazil. The first conclusion that this team reached in Germany was that the "underlying concepts in the sets of C&I evaluated were unclear and confusing." This was true even when the team consciously selected the sites that are considered to be much better than average with respect to having a perceived advantage for field testing C&I. CIFOR since then has been joined by others, such as the ITTO, the Tarapoto, the Lepaterique and the pan-European Processes to define or embark on defining C&I at the forest management unit level to complement those identified at the national and regional levels (see Interagency Partnership on Forests: Implementation of IPF Proposals for Action by ITFF, June, 1997; also see an online journal that CIFOR is making available on the World Wide Web in the name of C&I Updates (http://www.cgiar.org/cifor/research/C&I_info/whatsnew.html).

Some nongovernmental groups see the C&I process as too weak, too vague, and possibly a cover for legitimizing *business as usual* that permits unsound forest exploitation. A statement by an ad-hoc NGO Working Group on Forests sounds the alarm (NGO Statement on Forests to the Third Session of the U.N. Commission on Sustainable Development; a copy is available from the authors of this chapter):

> "But tragically, governments are on the wrong course. We do not support the ongoing governmental process of "harmonization" towards a global set of criteria and indicators (C&I) since, in our view: (a) global governmental C&I cannot account for the wide diversity of settings and conditions around the world; and (b) many governments intend to use a very general global set of C&I as the basis of a weak, trade-oriented forest convention, which we would oppose. More generally, the sharp focus on C&I has diverted precious time and resources from addressing the most important problems, the *underlying causes* of forest loss and degradation . . . We oppose the use of weak C&I as the basis for false "greenwash" certification of timber produced by business-as-usual environmentally destructive practices."

Several NGOs have formed themselves into the Forest Stewardship Council (Jamison Ervin, Interim Organizer, P.O. Box 849, Richmond VT 05477 USA) with the explicit objective of coming out with their own principles and criteria of forest management as a basis for timber certification.

Those who recommend the use of C&I as a starting point for drafting an international legal agreement claim that the C&I processes have had the imprimatur of the scientific community. In fact, these negotiating processes have been largely comprised of government representatives, perhaps with the input of a few governmental scientists, but certainly not with widespread comments of the scientific community. Nor do they necessarily offer a consensus among ecologists, foresters, and all the relevant scientific and technical disciplines regarding the conditions and processes by which sustainable forest management can be assessed. In general, the nongovernmental scientific community has not looked at the development of C&I principles from the perspectives of their own discipline. The objective of the symposium held at the 1995 annual meetings of the Soil Science Society of America and of this proceedings volume is to engage the attention of soil scientists to this end. It might be appropriate to study the same set of questions from the perspectives of other relevant disciplines.

NEW EVALUATIONS BY SOIL SCIENTISTS

Criteria and Indicators

We focus this book on the criteria and indicators developed by the Montreal Process. This is the only process that included countries with boreal, temperate, and tropical forests. The Montreal Process C&I were published as the Santiago Declaration (copies are available from the Canadian Forest Service, 351 St. Joseph Blvd., Hull, Quebec, Canada, K1A 1G5; the Santiago Declaration also was published in the Journal of Forestry, April, 1995). The criteria are listed in Table 1–1. Criterion 3 on forest ecosystem health and vitality and Criterion 5 on maintenance of forest contribution to global carbon cycles are relevant to soil resources. Indicators of these criteria include significant changes in soil nutrient cycling and carbon pools. Criterion 4 is the conservation and maintenance of soil and water resources, and the indicators for this criterion, given in Table 1–2, will be the focus of most of the discussion of forest soil scientists.

Table 1–1. List of criteria in the Santiago Declaration for the conservation and sustainable management of temperate and boreal forests.

1. Conservation of biological diversity
2. Maintenance of productive capacity of forest ecosystems
3. Maintenance of forest ecosystem health and vitality
4. Conservation and maintenance of soil and water resources
5. Maintenance of forest contribution to global carbon cycles
6. Maintenance and enhancement of long-term multiple socio-economic benefits to meet the needs of societies
7. Legal, institutional and economic framework for forest conservation and sustainable management.

Table 1–2. List of indicators in the Santiago Declaration for Criterion 4, the conservation and maintenance of soil and water resources.

1. Soil erosion: Percentage of forest area with significant soil erosion.
2. Protected areas: Percentage of forest land managed primarily for protective functions, e.g., watersheds, flood protection, avalanche protection, and riparian zones.
3. Stream flow: Percentage of stream kilometers in forested catchments in which stream flow and timing has significantly deviated from the historic range of variation.
4. Soil organic matter: Percentage of forest area with significantly low or diminishing soil organic matter.
5. Soil compaction: Percentage of forest area with significant compaction resulting from human activities.
6. Biological water quality: Percentage of water bodies (e.g., stream miles/kilometers, lake acres/hectares) with significant variance of biological diversity from the historic range of variability.
7. Chemical, nutrient, sediment, and temperature water pollution: Percentage of water bodies (e.g., stream miles/kilometers, lake acres/hectares) with significant variation from the historic range of variability in pH, dissolved O_2, levels of chemicals (electrical conductivity), sedimentation or temperature change.
8. Soil pollution: Percentage and area significantly affected by an accumulation of persistent or toxic substances such as heavy metals.

The Santiago Declaration defines C&I as the following:

- Criterion: "A category of conditions or processes by which sustainable forest management may be assessed."
- Indicator: "A measure (measurement) of an aspect of the criterion. A quantitative or qualitative variable which can be measured or described and which when observed periodically demonstrates trends."
- "A criterion is characterized by a set of related indicators which are monitored periodically to assess change."

Significance of Quantitative and Qualitative Indicators

The term *significant* stands out as a common key word in indicators shown in Table 1–2. This is problematic for scientists. Although threshold levels of extractable soil nutrients have been empirically derived for many agronomic crops (Mengel & Kirkby, 1987) and for some silvicultural applications (Binkley, 1986), an *a priori* value for the concentration of soil organic matter, for example, has never been suggested for sustainable management of forest soils, or any soils. Every textbook on introductory soil science emphasizes the beneficial effects of organic matter in soil, but none says how much is enough.

Scientists generally and soil scientists in particular are accustomed to working almost exclusively with quantitative data. Hence, terms like *soil quality* and *soil health* are met with considerable skepticism by many soil scientists (Allan et al., 1995). Yet scientists are being called upon to contribute to processes of assessing the status of forests, agricultural lands, and water resources within a complex social, economic, and political milieu. Some preliminary steps are being made in this respect for agricultural soils, where numerous soil properties have been measured and combined in an integrated analysis of changes in soil quality resulting from farm management practices (Allan et al., 1995; Reganold et al., 1993). The proposed criteria and indicators of sustainable conservation and management of forests are further examples of the need to integrate quantitative data with subjective judgements.

Quantitative assessments are the most straightforward and objective and are easiest to implement and to interpret. One of the roles for scientists in this process is to help identify quantifiable indicators that provide useful information on the status of forest soil resources. In some cases, however, potentially quantifiable indicators may be identified, but the few data that exist may be insufficient to provide meaningful interpretations. Historical data are needed if trends are to be identified, but few long-term studies of forest soil properties are available. Scientists will be needed to help determine, what, if anything, can be deduced from sparse data. In other cases, a variety of indicators may be available, perhaps each with a different degree of confidence in the data, and these indicators somehow must be integrated into an overall judgement of the status of the resource. Some degree of subjectivity is inevitable.

In one of the meetings leading up to the Santiago Declaration, the evolving role of criteria and indicators was discussed (Summary Conclusions of the Second Meeting of the Intergovernmental Working Group on Global Forests, October, 1994, Ottawa, Canada):

> "In general, indicators should be verifiable and quantifiable, but qualitative or descriptive indicators are also important. The issue of systematic observation and implementation requires substantial further work. Without the specification and commitment of resources to appropriate assessment systems, it will not be possible to evaluate whether forest management is moving towards or away from sustainability. Criteria and indicators will need to be reviewed and revised over time to reflect new research and improved understanding of forest management."

In short, the message is do the best job possible with the knowledge base and resources that are now available and continue to reassess as the knowledge base develops. The Ottawa meeting statement emphasized the developmental nature of the process of identifying scientifically based criteria and indicators:

> "The establishment of internationally agreed, scientifically based criteria and indicators should help provide a basis for:
>
> - assessment and evaluation of the progress made towards management, conservation and sustainable development of all forests at the local, national, regional and global levels;
> - promotion of international cooperation in management, conservation and sustainable development, of all forests;
> - formulation of national forest policies;
> - facilitation of international trade in forest products through support for development of timber certification schemes; and
> - common understanding of what is meant by the management, conservation and sustainable development of all types of forests world wide."

These are big stakes. Linking criteria and indicators to timber certification, in particular, means that they may affect an international market worth billions of dollars. With so much money involved, enforcement of the criteria and indicators could be manipulated to meet the needs of influential timber barons. As already noted, some environmental groups are suspicious that the criteria and indicators will be purposefully vague so that current logging practices can be carried on, business as usual, under the rubric of new internationally sanctioned agreements.

Even the official reports released by the Montreal Process have brought attention to the difficulties that many signatory countries have had in implementing the C&I and in providing data needed for the indicators. Shortcomings were particularly noted for Criterion 4 and its indicators, regarding forest soils (Progress on Implementation of the Montreal Process on Criteria and Indicators for the Conservation and Sustainable Management of Temperate and Boreal Forests, February, 1997; First Approximation Report of the Montreal Process, 31 Aug., 1997; Reports obtainable from Kathryn Buchanan, Liaison Office, Natural Resources Canada, Canadian Forest Service, Ottawa, Canada).

If C&I are used for timber certification and if they are vague, subjective, and open to variable interpretation, the potential for unfair and abusive application of C&I is large. On the other hand, if the criteria and indicators can be based on strong scientific underpinnings, linking them to timber certification could provide teeth for enforcement of improved forest management practices. It is the duty of scientists to provide that strong scientific basis or to identify infeasibility where the basis is too weak.

Questions for Soil Scientists

The importance of defining *significant* has already been emphasized. As a further point of departure, the following list of questions are appropriate for soil scientist to address:

1. Are the proposed indicators based on sound understanding of forest health in general, and on the known linkages between soil processes and forest health in particular?
2. Are the proposed indicators reliable measures of evaluating whether the criterion is being met?
3. Is measurement of the indicators feasible in most instances?
4. Where long-term records are not available, as is the case for most of the world, can short-term measures of the indicators reveal useful information about the criterion?
5. Where long-term records are available, would the application of these indicators reveal whether sustainable conservation and management has been achieved?
6. Are there potentially useful indicators of status of forest soils that could be measured and that are not listed?

CONTRIBUTIONS OF THIS VOLUME

Burger and Kelting start our discussion by providing a theoretical framework for how indicators of forest soil quality might be constructed from a first principals approach using current understanding of soil properties and ecosystem processes. They derive a forest soil quality index based on regionally specific indicators of soil properties appropriate for indicating sufficiency of root growth, water supply, nutrient supply, gas exchange, and biological activity. Unlike agricultural systems, where the productivity of this year's crop often directly reflects this year's soil conditions, growth of forests reflects long-term responses to variability in soils, climate, and management. These authors emphasize that changes in soil properties may affect long-term forest productivity in ways that short-term responses of tree growth may not reveal. Burger and Kelting then provide examples from field studies in the southeastern USA.

Powers, Tiarks, and Boyle present their experiences in studying changes in soil properties resulting from operational forestry in the USA. They propose three relatively simple measures that could be regionally calibrated as indicators of forest soil quality: soil strength measured with a pocket penetrometer, available nitrogen measured in an anaerobic laboratory assay, and field observations of soil invertebrate activity.

The effects of acid deposition and other forms of air pollution on forest health have been subjects of concern in Canada and Europe for over two decades. Morrison reviews early results from a long-term study of forest soil properties based on monitoring plots established throughout Canada to evaluate effects of acid deposition. Even when historical data such as these exist as a basis of comparison for current values of soil properties, issues of scale, confounding geographic factors, and within-plot spatial heterogeneity make interpretation of the results difficult. As countries begin implementing plans to monitor indicators of forest soils, the experiences of the few monitoring efforts now underway will be valuable.

European policy is based on the concept of critical loads of atmospheric inputs, reviewed here by Van Breemen, De Vries, and De Visser. The high growth rates observed recently in European forests may belie changes occurring in the soils that are cause for concern regarding future forest health, as well as the current condition of nontimber attributes of European forests. Defining critical loads of acceptable pollutants in Europe has depended upon process-level understanding of forest biogeochemistry as much or more than upon empirical relationships between doses and tree growth responses in the field. This concept should be examined as criteria and indicators for monitoring forest health are derived from inadequate data for establishing quantitative relationships between indicators and forest productivity, and instead, will initially rely on more basic understanding of the principals of how forest soil properties affect forest productivity, as clearly summarized here by Burger and Kelting and by Powers et al.

The Russian forest is vast, encompassing many climatic regions across Europe and Asia. Perhaps this helps explain the emphasis of Gorbachev and Laletin on mapping forests, soils, relief, climate, and others, in order to stratify this tremendous diversity into ecological zones that are relevant to management options and limitations. They argue that current forest extraction practices on highly erodible soils could and should be avoided if this type of zoning were developed. A system of monitoring criteria and indicators of forest soils would be important for both enforcement and for refining the mapping and zoning process.

In contrast to Russia, the native forests and plantation forests of New Zealand have been nearly completely delineated, with the former managed for nontimber attributes and the latter primarily for timber. Unlike the other regions described in previous chapters, acid deposition is not a major problem. Smith and Raison focus on how best management practices in operational forestry in Australia and New Zealand can be reinforced by the criteria and indicators process (and vice versa), although it will require new research to improve the definitions and calibrations of the indicators at the local scale, a theme echoed in every chapter.

Foresters may be surprised by the inclusion of a chapter on previously forested soils of Indonesia considered by Van Noordwijk, Hairiah, Woomer, and Murdiyarso. Agroforestry systems in tropical regions blur the distinction between forest and agricultural soils, but a myriad of land uses with varying degrees of cover by tree species is evolving in tropical regions throughout the world. These authors take advantage of an unusual data set on soil C contents in the island of Sumatra measured in the 1930s and remeasured in the 1980s, with extensive land use change occurring in the interim. They derive relationships between soil C content, texture, acidity, soil type, and elevation to estimate a reference value for expected forest soil C content. The effects of current land uses on soil C content can then be evaluated based on the relationship between current C content and its reference value. They also propose a soil C fractionation procedure, that is relatively simple, but will require further standardization.

CONCLUSION

If the criteria and indicators are too weak or are unworkable, the onus is on the independent scientific community to say so now. If they can be improved, these suggestions would improve policy formulation with the best science now available. It will not be enough to say that more study is needed, although that conclusion is almost always true. The available data will no doubt be incomplete, and many scientists may feel uncomfortable making recommendations without the security of a 95% confidence interval. Some subjectivity will be necessary. These are not the rules under which most scientific papers are published, but these are the conditions under which we are now asked to contribute to a process of developing good policy.

Negotiators cannot be the sole arbiters of what constitutes appropriate criteria and indicators if a workable and effective policy is to be achieved. It would be unfortunate if forest policies are established without input from scientists outside the government sector. If science is to lead policy, as most of us believe it should, then forest soil scientists must take the opportunity now presenting itself to evaluate the proposed criteria and indicators of sustainable forest conservation and management.

ACKNOWLEDGMENTS

Support for preparation of this article was provided by a grant from the John D. and Catherine T. MacArthur Foundation to the Woods Hole Research Center.

REFERENCES

Allan, D.L., D.C. Adriano, D.F. Bezdicek, R.G. Cline, D.C. Coleman, J.W. Doran, J. Haberern, R.F. Harris, A.S.R. Juo, M.J. Mausbach, G.A. Peterson, G.E. Schuman, M.J. Singer, and D.L. Karlen. 1995. SSSA statement on soil quality. *Agronomy News*, June, p.7.

Binkley, D. 1986. Forest nutrition management. John Wiley & Sons, New York.

Food and Agriculture Organization. 1994. The state of food and agriculture, 1994. FAO, Rome.

Houghton, R.A. 1993. Role of the world's forests in global warming. p. 21–58. *In* K. Ramakrishna and G.M. Woodwell (ed.) World forests for the future: Their use and conservation. Yale Univ. Press, New Haven, CT.

Maini, J.S., and O. Ullsten. 1993. Conservation and sustainable development of forests globally: Issues and opportunities. p. 111–119. *In* K. Ramakrishna and G.M. Woodwell (ed.) World forests for the future: Their use and conservation. Yale Univ. Press, New Haven, CT.

Mengel, K., and E.A. Kirkby. 1987. Principles of plant nutrition. 4th ed. Int. Potash Inst., Bern, Switzerland.

Oksanen, T., M. Heering, and B. Cabarle. 1993. A study on coordination in sustainable forestry development. Int. Inst. for Environment and Development, Street, London.

Prabhu, R., J.P. C.J.P. Colfer, P. Venkateswarlu, L. Tan, R. Soekmadi, and E. Wollenberg. 1996. Testing criteria and indicators for the sustainable management of forests: Phase 1. Final Report. Ctr. for Int. Forestry Res. Spec. Publ. Ctr. for Int. Forestry Res., Jakarta, Indonesia.

Ramakrishna, K. 1993. The need for an international commission on the conservation and use of world forests. p. 121–134. *In* K. Ramakrishna and G.M. Woodwell (ed.) World forests for the future: Their use and conservation. Yale Univ. Press, New Haven, CT.

Reganold, J.P., A.S. Palmer, J.C. Lockhart, and A.N. Macgregor. 1993. Soil quality and financial performance of biodynamic and conventional farms in New Zealand. Science (Washington, DC) 260:344–349.

Repetto, R. 1993. Government policy, economics, and the forest sector. p. 93–110. *In* K. Ramakrishna and G.M. Woodwell (ed.) World forests for the future: Their use and conservation. Yale Univ. Press, New Haven, CT.

United Nations. 1992. Report of the United Nations conference on environment and development. UNDoc. A/CONF.151/26, 14 Aug. 1992. United Nations, New York.

United Nations. 1995a. Report of the Secretary General to the United Nations Commission on Sustainable Development, Land, Desertification, Forests and Biodiversity. UNDoc. E/CN.17/1995/3, 14 Feb. 1995. United Nations, New York.

United Nations. 1995b. Regional workshop on the definition of criteria and indicators for sustainability of Amazonian forests. UNDoc. E/CN.17/1995/34, 10 Apr. 1995. United Nations, New York.

United Nations. 1997a. Report of the ad hoc Intergovernmental Panel on Forests on its fourth session, UN Doc. E/CN.17/1997/12, 20 Mar. 1997. United Nations, New York.

United Nations. 1997b. Report of the Intergovernmental Forum on Forests on its first session, UN Doc. E/CN.17/IFF/1997/4, 10 Oct. 1997. United Nations, New York.

Woodwell, G.M. 1993. Forests: What in the world are they for? p. 1–20. *In* K. Ramakrishna and G.M. Woodwell (ed.) World forests for the future: Their use and conservation. Yale Univ. Press, New Haven, CT.

2 Soil Quality Monitoring for Assessing Sustainable Forest Management

James A. Burger and Daniel L. Kelting

Virginia Polytechnic Institute and State University
Blacksburg, Virginia

The notion of managing forest, land, and water resources sustainably has been advocated for at least a century by many forest managers, landowners, policy makers, and philosophers, but exactly what should be sustained, for whom, for how long, and under what conditions have been vigorously debated to the present. In the USA, a number of federal laws, such as the Multiple Use and Sustained Yield Act of 1964, the Surface Mining Control and Reclamation Act of 1977, and the Clean Water Act of 1989, resulted from these national debates, and their implementation has had a positive long-term influence on the quality of forests, soil, water, and land. Nevertheless, the debates continue as various stakeholders argue about the role and place of biology, economics, and society's needs in sustainable development as it pertains to forests and forestlands. Recent debates have shown that the conservation ethic, "use of resources for the greatest good, for the greatest number, for the longest time," that provided the basis for resource conservation for the past 75 yr, is inadequate for ensuring sustainable development. Beginning with the publication of the Brundtland Report (World Commission on Environment and Development, 1987) and followed by the 1992 United Nations Conference on Environment and Development (Earth Summit; Ramakrishna & Davidson, 1998, this publication), the conservation ethic as it was applied to forests and forestry has been replaced by a sustainability ethic as the guiding principle for natural resource management. The sustainability ethic is centered less on consumption of resources and more on the idea of perpetual resource availability. Unlike the conservation ethic, it explicitly addresses the notion of intergenerational equity.

DEFINING SUSTAINABLE FOREST MANAGEMENT

Despite wide acceptance of the concept of sustainability, applying it to forests and forestry has been a challenge. Sustainable forestry has yet to be precisely defined. In 1992, the Wilderness Society's Bolle Center for Forest

The Contribution of Soil Science to the Development of and Implementation of Criteria and Indicators of Sustainable Forest Management. SSSA Special Publication no. 53.

Ecosystem Management, American Forests' Forest Policy Center, and the World Resources Institute convened a national conference to explore our collective understanding of sustainable forestry (Aplet et al., 1993). A recurrent theme during this national conference was that there is no one perfect definition of sustainable forestry. Because forestry is a human construct, defining sustainable forestry requires resolving what societies want from their forests (Aplet et al., 1993). Therefore, a definition of sustainable forestry will usually reflect "the eye beholding the forest."

The American Forest and Paper Association, the national trade group that represents forest and paper companies in the USA, is one of very few organizations committing to a definition (AFPA, 1995):

> "*Sustainable forestry means managing our forests to meet the needs of the present without compromising the ability of future generations to meet their own needs by practicing a land stewardship ethic which integrates the growing, nurturing and harvesting of trees for useful products with the conservation of soil, air and water quality, and wildlife and fish habitat.*"

This definition provides for intergenerational equity and environmental protection of adjacent resources, while producing forest products to meet the needs of the present generation. This definition contains the major themes found in most discussions except for "economic viability of forestry business," which is surprising, considering the source.

While definitions vary, there is fairly widespread agreement on what sustainable forestry should be: It must simultaneously be ecologically viable (environmentally sound), economically feasible (affordable), and socially desirable (politically acceptable; Salwasser et al., 1993). These attributes were captured by the Helsinki Process (1994), formal discussions of a group of representatives from 36 European countries charged with developing general guidelines for the sustainable management of forests in Europe. This group defined sustainable forest management as:

> "*. . . the stewardship and use of forests and forest lands in a way and at a rate, that maintains their biodiversity, productivity, regeneration capacity, vitality, and their potential to fulfill, now and in the future, relevant ecological, economic, and social functions, at local, national, and global levels, and that does not cause damage to other ecosystems.*"

This sustainability ethic is being considered, and in some cases applied, within the global forestry community with the idea of sustaining in perpetuity the forest ecosystem, the business of forestry, and human communities that depend on forests. With the momentum of the Rio Earth Summit, five multi-country United Nations Conference on Environment and Development (UNCED) initiatives ensued on developing criteria and indicators of sustainable forest management at national levels (Ramakrishna & Davidson, 1998, this publication). All five initiatives include aspects of seven basic criteria: (i) forest health and vitality; (ii) extent of forest resources; (iii) maintenance of productive functions; (iv) biological diversity; (v) protective and environmental functions; (vi) developmental and social needs; and (vii) legal, policy, and institutional frameworks. Presumably, if all criteria are simultaneously maintained, then the biology of forests, the business of forestry, and forest-based human communities should be sustained in perpetuity to serve present and future generations. These initiatives cover the majority of the world's forests and, with time, are sure to have a significant impact. During the rest of this decade, discussions at national levels will no doubt continue in order to determine how to implement these criteria at various scales, and to determine whether or not sustainable forest management is being achieved.

SOIL AS A CRITERION OF SUSTAINABLE FOREST MANAGEMENT

The emphasis of this chapter is on the role of soils in sustainable forest management. Across the earth's arable landscapes, human generations have had a profound effect on soils and their productivity; soils of many landscapes have been degraded (Oldeman, 1994; Lal & Pierce, 1991). The greatest threat to forest soils comes with the conversion of forestland to pastureland or agriculture; however, forest soil productivity in managed forests also may be threatened by certain management practices (Dyck & Skinner, 1990; Ballard & Gessel, 1983; Johnson, 1994). Because soil loss, like species loss, is irreversible in a human time frame (Franklin, 1993; Burger, 1997), soil conservation ranks with biodiversity as a sustainability criterion requiring immediate attention.

The influence of soils on forests is reflected in several criteria common to the five multi-country initiatives, such as the Montreal Process, including "maintenance of forest health and vitality," "maintaining a balanced carbon cycle," and "maintaining protective and environmental functions." Soil-based indicators, such as soil erosion, diminished organic matter, and compaction are proposed in the Montreal Process (Ramakrishna & Davidson, 1998, this publication). In a general sense, these are important determinants of soil productivity and function as argued by Powers et al. (1998, this publication), but at this time there is no suitable applications framework for their use as measures of sustainable forest management.

During March, 1996, a group of forest resource professionals and technical experts met in Washington, DC, for a Criteria and Indicators Workshop to assess, from a U.S. perspective, the seven criteria and indicators of sustainable forest management set forth by the Montreal Process, of which the USA was a participant. Regarding soil-based indicators, there was general agreement among participants that these indicators may have some use at the national level for broad policy decisions, but that data collection and interpretation would be difficult. The committee noted specifically that (i) terms need to be defined (e.g., compaction, significant), (ii) that standard methods for measuring soil properties are needed, (iii) site specificity of indicators is required, (iv) indicators must be connected to some component of forest function, (v) baselines are needed to put measures in perspective, and (vi) the scale at which criteria and indicators are being applied must be clarified.

Clearly, much was accomplished by the multi-country initiatives to develop criteria and indicators for sustainable forest management, but the work has been largely conceptual and was done at a very broad scale. Much additional work on all criteria is needed to further define indicators and develop mechanisms for implementing monitoring systems at various scales for specific purposes. The purpose of this chapter is to clarify the rationale and outline an approach for the selection and use of soil-based indicators to help monitor forest management to ensure that it is and remains sustainable.

RATIONALE FOR USING SOIL-BASED INDICATORS TO ASSESS SUSTAINABILITY

We believe an overall goal of a soil-based monitoring system for assessing sustainable forest management is to ensure that forest soils, when exposed to various forest management practices, retain or improve their capacity to (i) sustain plant and animal productivity, health and vitality, (ii) maintain balanced hydrologic, carbon, and mineral nutrient cycles, and (iii) maintain protective and environmental forest functions, all of which are criteria common to the multi-country initiatives in which forest soils play a role.

Through a monitoring system, we should address specific questions about a soil's condition to determine whether or not soils are functioning in a way that will allow us to meet our goals and objectives. For example, the question implicit in Criterion 4 of the Montreal Process is: "Are soil and water resources being conserved and maintained?" Other questions that a soil-based monitoring system should address: (i) Is soil tilth optimal and being maintained for plant growth? (ii) Is the soil in the best physical condition to accept, hold, and regulate water in ways that ensure its quality and optimize its delivery rate to plants and off-site aquatic systems? (iii) Is the soil biological community and its abiotic environment in a condition that maximizes the renovation of applied waste materials with no off-site effects? (iv) Is the soil system serving as a net sink for C and storing C commensurate with its natural capacity? (v) Is the soil storing, supplying, and cycling mineral nutrients at levels and rates that optimize plant productivity? Furthermore, monitoring these questions must be specific for (i) soil *functions* (productivity, environmental quality, C cycling, and others), (ii) various spatial *scales*, and (iii) different *forest management systems* (natural stands, plantations, short-rotation woody crops, and others).

Function

Promoting plant and animal productivity, regulation of the hydrologic cycle, maintenance of global C cycles, and serving as a medium for the recycling of organic wastes are several distinct *functions* of forest soils. One or more of these functions may be emphasized on a given land area. Extent of soil erosion, compaction, and soil organic matter levels, indicators for Criterion 4 from the Montreal Process, may be appropriate for monitoring some functions, but may not be adequate for others. Indicators of sustainable forest management should be selected and targeted to measure each function considered important. Some potential indicators such as soil water-retention capacity and soil organic matter content may measure several functions simultaneously, while others may be function-specific. The function of promoting forest productivity at the stand level will be used in examples to illustrate other aspects of choosing and using criteria and indicators without losing sight of the fact that indicators for each soil function are needed.

Spatial Scale

Another aspect of accomplishing sustainable forest management is to determine the scale at which a monitoring program will be applied. According to background documents accompanying the Santiago Declaration, the culmination of the Montreal Process, the criteria and indicators of sustainable forest management are intended to evaluate a country's progress toward sustainability at the national level. They are not intended to directly assess sustainability at the forest management unit level. If their validity is established, they will be put to use at national and international policy-making levels; however, their validity will depend upon the success of using criteria and indicators at smaller scales. Levels of forest management scale are hierarchical, so forest system sustainability at any level can only be achieved when sustainability at the next lower level is achieved (Table 2–1). Four levels of hierarchical scale, landscape, forest, site, and soil, are shown in Table 2–1 for evaluating sustainable forest productivity. Much like erecting a building, the foundation must first be laid (soil level), followed by the floor (site level), walls (forest level), and ceiling (landscape level) before a roof (national level) is added and becomes useful. And the materials of construction are level-specific. Indicators for measuring and monitoring sustainability appropriate at one level may not be useful at a higher or lower level. For example, organic matter content, soil density, soil fertility, and soil water holding capacity may vary greatly at the site level between a managed natural forest and a short-rotation woody-crop plantation. Two different levels of organic matter content could be appropriate for the two systems, but this fact may not be recognized in a broad national-level assessment. Furthermore, the use of a direct measure of productivity reflecting management-induced soil change (Table 2–1, Column 4) vs. a measure of productive capacity (Table 2–1, Column 5) also will be scale-specific. The extent and degree of soil erosion and compaction and organic matter depletion, indicators suggested for the national level in the Santiago Declaration, are more appropriately assessed at the soil and site level.

Forest Management System

The business of forestry in temperate and boreal forests is conducted in managed natural forests, long-rotation plantation forests, and short-rotation woody crop systems (SRWC; Fig. 2–1). Among these managed forests, there exists a large gradient in system complexity. Sustainable management might be achieved in natural forests by encouraging natural processes and maintaining the forest within the historical range of conditions observed for a particular seral stage (Kimmins, 1996). To manage woody crops sustainably, on the other end of the management-intensity spectrum (Fig. 2–1), we might need to apply some management and soil conservation techniques used in agriculture. Long-rotation

Table 2–1. Characterization of scale for applying soil-based criteria and indicators of productivity for achieving sustainable forest management.

Sustainability level and components at each level	Spatial scale	Management level	Direct measures of productivity	Measures of productive capacity
	ha			
Landscape: • human communities • hydrologic cycle • forest	1000–100	Forest region or watershed	Sum of forest products and services	Level of soil functioning for maintaining forest health, water quality, and human welfare
Forest: • vegetation • habitat • site	100–10	Multiple stands	Volume of canopy trees and density–diversity of wildlife	Level of nutrient cycling, hydrologic cycling, and energy transfer
Site: • climate • physiography • soil	10–1	Forest stand	Height of canopy trees	Soil volume, fertility, air– water balance
Soil: • physical properties • chemical properties • biological properties	1–0.1	Substand	Aboveground NPP minus herbivory	Soil properties and processes

plantation forests, intermediate in intensity, will need to be managed with a combination of approaches. The amount of cultural input and soil manipulation varies greatly among these types of forest management systems. Sustainability criteria may be the same, but the indicators that measure success in applying the criteria must be different because the systems are very different. An integrative approach using biomass production or some other vegetation performance indicator (e.g., production assays) may be adequate for monitoring the sustainability of managed natural forests, while a more reductionist approach using soil condition indicators (e.g., soil quality characteristics) may be best suited for assessing the sustainability of short-rotation woody crops.

Need for Soil-Based Indicators

Soil is the medium containing the below-ground structures of trees, and we recognize that soil plays important roles in forest biogeochemical and hydrologic cycles and energy transfers in forest ecosystems; however, unlike annual-crop agriculture, where soil productivity is studied, managed, and nurtured apart from the crop grown, foresters have traditionally treated *soil* productivity as a component of overall *forest* productivity, especially in the managed, natural, temperate, and boreal forests (Fig. 2–1). The influence of soil on forest productivity is seldom separated from the influence of other abiotic site factors such as climate, geology, and physiography. Where species compositions and soil physical, chemical, and biological properties are not significantly manipulated, a bioassay measure such as species diversity, relative growth, biomass increment, or height (site index) that integrates the influences of all growth, health, and vitality determinants may suffice as an indicator of sustainable forest management at the site level (Table 2–1), with no need of a more reductionist measure at the soil level.

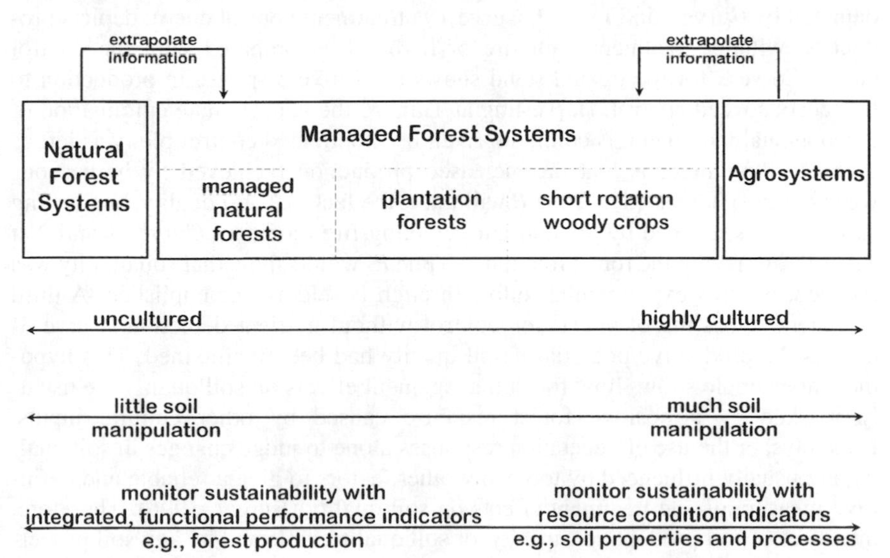

Fig. 2–1. Spectrum of forest management types for which sustainability monitoring systems are needed (after Burger, 1997).

For example, if harvesting simply converts a forest ecosystem condition from one seral stage to another without impairing the processes of recovery (similar to natural disturbances), then forest integrity could be monitored at the site and forest levels using vegetation-based indicators (Table 2–1, Column 4).

In plantations and short-rotation woody crop systems, combinations of clearcutting, burning, drainage, soil tillage, fertilization, and planting with genetically improved stock are common. Neither biotic potential (genotype) nor site carrying capacity (soil productivity) are fixed (Stone, 1975). Forest stand productivity is influenced by improving genotypes, by changing soil quality, and by shifting ephemeral resources from subordinate vegetation to crop trees. Measuring and monitoring sustained forest productivity is more complex in plantation forests because all determinants of production are manipulated by management. The influence of forest practices on forest stand productivity is often a function of a combination of changes in species composition, plant density, plant interactions, and soil treatments, but the separate influence of each on volume yields is seldom known. Production comparisons of successive rotations at the forest or site level are considered the ultimate measure of long-term productivity by most forestry researchers (Morris & Miller, 1994); however, differences in production between growth cycles are a function of so many factors (biotic, abiotic, cultural) that causation, or reasons for production differences between growth cycles can never be determined definitively (Burger, 1996; Ford, 1983).

The effects of any single silvicultural treatment on plantation stand growth are fairly well understood. But an understanding of cumulative response to multiple treatments over time on sites of different character is much more difficult. For example, a pine plantation established by a hot site preparation burn used to control residual hardwoods and clear the site for planting, followed by herbaceous weed control one year after planting, may result in a production curve depicted by Curve 2 in Fig. 2–2. Curve 1, a treatment control curve, depicts production without treatment with fire or herbicide. Compared with the control curve, Curve 2 for the treated stand shows a positive response in production to herbaceous weed control. Harvesting at Time *R*, the approximate culmination of mean annual increment, is common. Even if a nontreated control plot was left as a check, the danger is that the increased production (achieved by herbaceous weed control) harvested at Time *R* will mask the fact that soil quality, or carrying capacity, was reduced by the nutrient-depleting fire (compare Curves 1 and 2 at Time *R′*). Carrying the forest rotation to Time *R′* would show that soil quality was reduced, but this experimental follow-through is seldom accomplished. A third treatment consisting of vegetation control without nutrient depletion (Curve 3) depicts the productive potential if soil quality had been maintained. This hypothetical example shows how forest management effects on soil quality are usually masked by a positive forest response caused by other cultural inputs. Bioassays, or the use of vegetation responses alone to judge changes in soil quality, are usually influenced by too many other factors to give a reliable and definitive measure of site treatment effects on soil quality (Burger, 1996). Therefore, measures of soil productive capacity, or soil quality, reflected by key soil properties and processes (soil attributes), are necessary for assessing sustainable forest management in forest systems where soils are directly or indirectly manipulated.

There is a significant move afoot in the agricultural community to develop soil quality monitoring programs to ensure that agricultural production, environmental quality, and human and animal health are sustained (Doran et al., 1994; Doran et al., 1996; Papendick & Parr, 1992). In spite of the fact that sustainability of agricultural practice might be monitored directly using year-to-year production data, attempts are being made to monitor agricultural sustainability indirectly by integrating soil parameters into a meaningful index that correlates with soil productivity, environmental quality, and animal health (Parr et al., 1992). Indirect estimates of agricultural soil quality were also motivated in part by the inability of assessing management effects on soil productivity apart from effects of plant breeding and other cultural inputs. During most of this century, yields from agricultural crops increased due to improved genotypes and intensive culture despite the fact that soil quality was deteriorating, ultimately causing yields to level off and decline in some cases (Lal & Pierce, 1991). Therefore, efforts to monitor soil quality apart from crop productivity have become necessary.

The role of forest soil scientists in achieving sustainable forest management is to identify, test, and validate methods for monitoring soil quality and forest productivity at all levels, and participate with foresters and land managers to ensure that each level of management is adequately underpinned with appropriate, scientifically-based knowledge. Systems for monitoring agricultural sustainability may serve as models for developing methods for monitoring forest sustainability, but Nambiar (1996) cautions that there are several important differences between the two systems. Properties and processes of agricultural soils reach a quasi-equilibrium; deviations from this equilibrium are relatively easy to detect and the

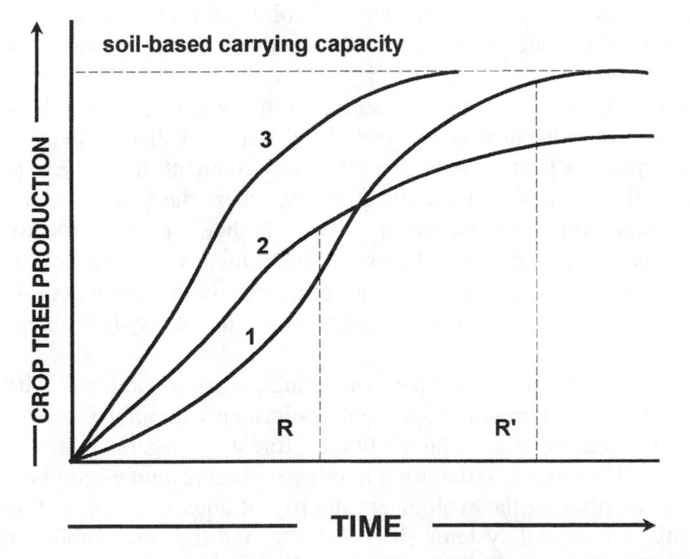

Fig. 2–2. Hypothetical production curves: Curve 1 depicts a treatment control curve; Curve 2 depicts a treatment that reallocates resources to desired species but reduces site carrying capacity; Curve 3 depicts a treatment that reallocates resources to desired species while maintaining site carrying capacity. *R* represents rotation length at the culmination of economic mean annual increment; *R*' represents stand age when maximum production is realized (Burger, 1994).

causal agent identified. Soils under longer-lived forests fluctuate with time as they are influenced by stage of forest growth and natural recovery from management perturbations. Therefore, point-in-time indicators must be carefully chosen to reflect long-term effects.

The older and more complex the forest, the more likely the predictive capacity of indicators will change as the age and structure of the forests change (Nambiar, 1996). With short-rotation forest systems, where soil productivity is less a function of vegetation influences and multiple interacting factors (the agricultural model), soil-based indicators, if properly validated, should be useful for assessing sustainability.

A FRAMEWORK FOR MONITORING FOREST SOIL QUALITY TO ACHIEVE SUSTAINABLE FOREST MANAGEMENT

The concept, definition, and general principles for *soil quality* have been well developed and rationalized by agricultural soil scientists. Karlen et al. (1997) define soil quality and present the history of its development as a concept. Foresters should readily identify with it because in principle, it is very similar to *site quality*, a concept that has been used for decades. Site quality, the sum of the effect of all site factors on the capacity of a forest site to produce plant biomass, includes site factors such as climate, elevation, slope, aspect, landscape position, and geologic influences as well as soil factors. It is usually indexed with *site index*, the height of canopy trees at a given age (Carmean, 1975). Site index has been very useful historically for characterizing and delineating the natural site carrying capacity or productive capacity of abiotic site factors; however, it does not adequately capture management-induced changes in soil productivity (Monserud, 1984).

Soil quality is an expression of soil attributes within fixed levels of site factors. It is the sum of the effects of selected soil attributes that are the most important determinants of plant growth in a given environment. It can be expressed as an index (SQI) of a soil's productive capacity, where the index is some fraction of that of a soil with no limitations on growth. It should be more useful than site index for measuring management or other human-caused effects on forest productivity, since forest practices and other human activities influence soil quality, but rarely do they influence other components of site quality (e.g., slope, aspect, and others).

We suggest that a soil quality monitoring program combined with process models and traditional growth response measurements should be used to monitor soil change to ensure the sustainability of forest management practices. The approach could be similar to that used in other natural resource-based disciplines. Systems for monitoring the ecological integrity of aquatic systems (Karr, 1993), biodiversity of terrestial systems (Noss, 1990), and the sustainability of agroecosystems (Doran et al., 1994) have been devised. After the goals, objectives, and endpoints have been determined, a number of steps are common to these monitoring systems: (i) gather the scientific knowledge of the system, however incomplete, and use it to underpin all aspects of a monitoring system, (ii) select indicators that will serve as measurable surrogates of soil attributes that define soil quality, (iii) establish baseline conditions against which to compare conditions of the managed system, (iv) monitor all management practices that cause significant changes in soil properties, (v) implement a sampling scheme for measuring indicators across space and time, (vi) validate relationships between indicators and goals, (vii) analyze trends and recommend management actions, and (viii) employ adaptive management. Refine and change guidelines and practices as needed.

Scientifically-Based Monitoring

Good monitoring systems are nearly always based on a body of scientific knowledge. The scientific knowledge base for forests, forestry and soils is both broad and deep and is readily available in excellent forestry and soil science journals. Specific science topics germaine to sustainable forestry have been further summarized in many published reviews and symposium proceedings. Examples of recent excellent reviews on forest soil productivity and soil quality that summarize the literature on which to base a system for monitoring sustainability include those edited by Lal and Stewart (1995), Greenland and Szabolcs (1994), Dyck et al. (1994), Doran et al. (1994), and Gessel et al. (1990). While humbly recognizing that there is much unknown about forest and soil systems, it is clear, nonetheless, that there is a vast body of literature to draw upon for developing science-based monitoring systems.

A simple literature-based conceptualization of the components of forest soil structure and function on which soil quality and forest productivity depend is shown in Fig. 2–3. The role of soils in forest productivity is primarily a function of availability and supply of water, air, and nutrients synchronized to meet the

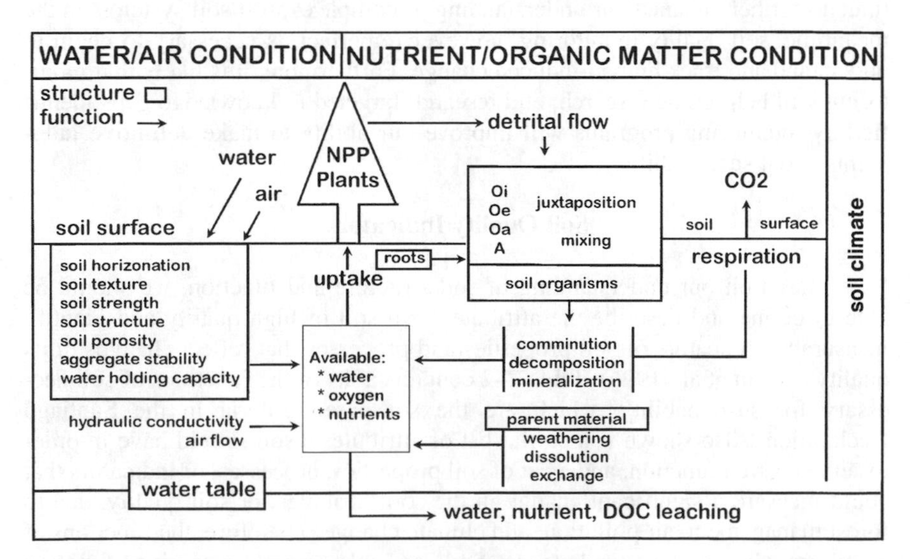

Fig. 2–3. Outline of components of forest soil structure and function.

diurnal, seasonal, and phenological demands made by individual plants and groups of interacting plants. Soil, water, air, and nutrient supply is a function of a combination of soil physical, chemical, and biological properties and processes that can be grouped into two resource regions (Fig. 2–3) that control water–air condition and nutrient–organic matter condition. In the diagram of Fig. 2–3, soil properties are depicted as structural components in boxes, and soil processes are depicted as arrows showing flow of materials from one system component to another. Soil water and O_2 supply is controlled by structural properties, including soil texture, structure, horizonation, aggregate stability, and water holding capacity, and by functional properties, including hydraulic conductivity, air flow, and water table flux. Nutrient supply is controlled by structural properties, including presence and arrangement of detrital material, soil organisms, and soil parent materials, and by functional processes, including comminution, decomposition, mineralization, weathering, dissolution, and ion exchange. Water, nutrients, air, and organic matter move into and out of the soil at certain rates and times partly due to soil properties and processes, and partly due to factors outside the influence of soil.

Therefore, a forest soil is a complex body with well-established properties and processes that interact to cause and allow water, air, and nutrients to cycle and energy to be transferred in ways that promote plant productivity. If our understanding of the soil system were complete, we could predict precisely how forest practices affected the soil's productive capacity. But since our understanding is only partial, monitoring change in these key properties and processes through time, whose function and role in sustainability we understand, is the best available way of assessing soil-based criteria of sustainable forest management. The more complete our scientific knowledge of the soil system, and the more knowledge incorporated into a monitoring system, the better we will be able to judge whether forest management is sustainable. Basic and applied research must continue to further enhance our understanding of complex forest soil systems. In the meantime, soil quality monitoring, however imperfect, is necessary to begin to understand and track human-induced change. Furthermore, it is likely that monitoring will help guide research, and research targeted to knowledge gaps identified by monitoring programs will improve our ability to make definitive judgments about sustainability.

Soil Quality Indicators

Based on our understanding of soil structure and function, we should be able to define and describe the attributes of a soil of high quality and identify measurable indicators of soil properties and processes that reflect changes in its quality (Doran et al., 1996). Table 2–2 contains a list of the functions of soil necessary for sustainability, which are the soil-based criteria in the Santiago Declaration. Also shown is a partial list of attributes a soil should have in order to serve a given function, and a list of soil properties, processes, or indicators that could measure a significant change in the soil attributes, or soil quality, due to forest management, air pollution, and climate change. Therefore, the functions of a soil are what we want soils to do. Soils are able to perform a given function

Table 2–2. Partial list of forest soil functions, attributes, and indicators for assessing sustainable forest management.

Soil function	Soil attributes	Indicators, indicator constructs: Level I	Indicators, indicator constructs: Level II	Values–units
Forest productivity	promote root growth	soil strength	penetrometer force	MPa–depth
		least-limiting water range (de Silva et al., 1994)	volumetric water BD, OM, structure	% by volume
		soil tilth index (Singh et al., 1992)	BD, strength, aggregate uniformity, SOM, plasticity index	index 0–1
	accept, hold, and supply water	infiltration		cm H_2O min^{-1}
		water holding capacity	Θ vol. between 1/3 bar–15 bar vol.	cm H_2O cm^{-1} soil
		unsaturated hydraulic conductivity	soil texture	% silt and clay
		water table depth		cm
	hold, supply, and cycle nutrients	N supply	N mineralization	kg ha^{-1} yr^{-1}
		OM content	C content	kg ha^{-1}
			soil respiration	Mg CO_2 ha^{-1} yr^{-1}
			active organic matter	C active/C total
		CEC	effective CEC	C mol kg^{-1}
		labile nutrient content	extractable nutrient	kg ha^{-1}
		pH		pH units
	promote optimum gas exchange	porosity	macroporosity	% soil volume
		water content	redox potential	mV
		air content	O_2 level	ppm or %
	promote biological activity	biologically active OM content		kg ha^{-1}
		soil temperature		°C
		soil moisture		Θ vol
		pH		pH units
		microbial biomass		kg ha^{-1}
Regulate hydrologic cycle	accept, hold, and release water	infiltration		cm H_2O min^{-1}
		water holding capacity	macroporosity	
		evaporation	surface roughness	roughness index
Regulate C balance	accept, hold, and release C	litterfall		kg ha^{-1} yr^{-1}
		root turnover		kg ha^{-1} yr^{-1}
		soil respiration		Mg CO_2 ha^{-1} yr^{-1}
		SOM		kg ha^{-1}

based on the extent to which they possess certain attributes. By measuring the levels of soil attributes, we can judge the soil's quality or its ability to function. If the attributes cannot be measured directly, surrogate properties or indicators can be used to measure the condition of the soil attributes and monitor human-caused changes. Some soil attributes may be more difficult to measure than others, requiring two or more indicators.

To be most useful, soil quality attributes or their indicators should (i) have an available baseline against which to compare changes, (ii) provide a sensitive and timely measure of the soil's ability to function, (iii) be applicable over large areas, (iv) be capable of providing a continuous assessment, (v) be inexpensive and easy to use, collect, and calculate, (vi) discriminate between natural changes and those induced by management, and (vii) be highly correlated to long-term response in long-lived forest systems (Nambiar, 1996; Doran et al., 1996; Noss, 1990).

A pie chart model is used to illustrate the concept of soil quality and the features of indicators that measure sustainability (Fig. 2–4A). Soil quality is a function of many attributes, but, for a particular site, attributes or their indicators are chosen because they are expected to be correlated with forest productivity, and because they are sensitive to impacts that are imposed by management practices (hypothetical indicators A through G). They are soil attributes or indicators of attributes that are neither resistant nor resilient to change. When an impact on the soil occurs, the indicator responds and the change in the indicator remains as long as the impact remains.

Soil quality is indexed as a fraction of one (area of the circle in this spatial model). Some indicators may be more important than others, as shown by the weighting factor or angle of the wedge; the sum of the areas of the wedges is equal to one, or 100%, and is depicted as the area of the circle. The area of the circle represents the very best soil quality with no limitations on growth; it will support the highest level of productivity for a given forest type.

Figure 2–4B shows a site quality pie chart for a hypothetical forest site where soil quality is limiting productivity. The sum of the area of the wedges is equivalent to the area of the circle representing an overall productivity level of 0.60, or 60% of the maximum growth level. The example shows that some indicators are sufficient for maximum growth (sufficiency level 1.0), while others are insufficient. If the soil quality indicators have been properly selected and weighted, they can be targeted for improvement through management to increase soil quality. After management practices are applied, indicators can be remeasured and used to assess the change in forest soil quality (FSQ) by comparing the FSQ against the original baseline.

Soil Quality Baseline

A soil quality reference condition or baseline might best be represented as the natural level of selected soil attributes for a given region or site. For example, the conditions shown in Fig. 2–4B could be those of a natural undisturbed soil prior to any manipulation with forest practices. This hypothetical soil is shown to

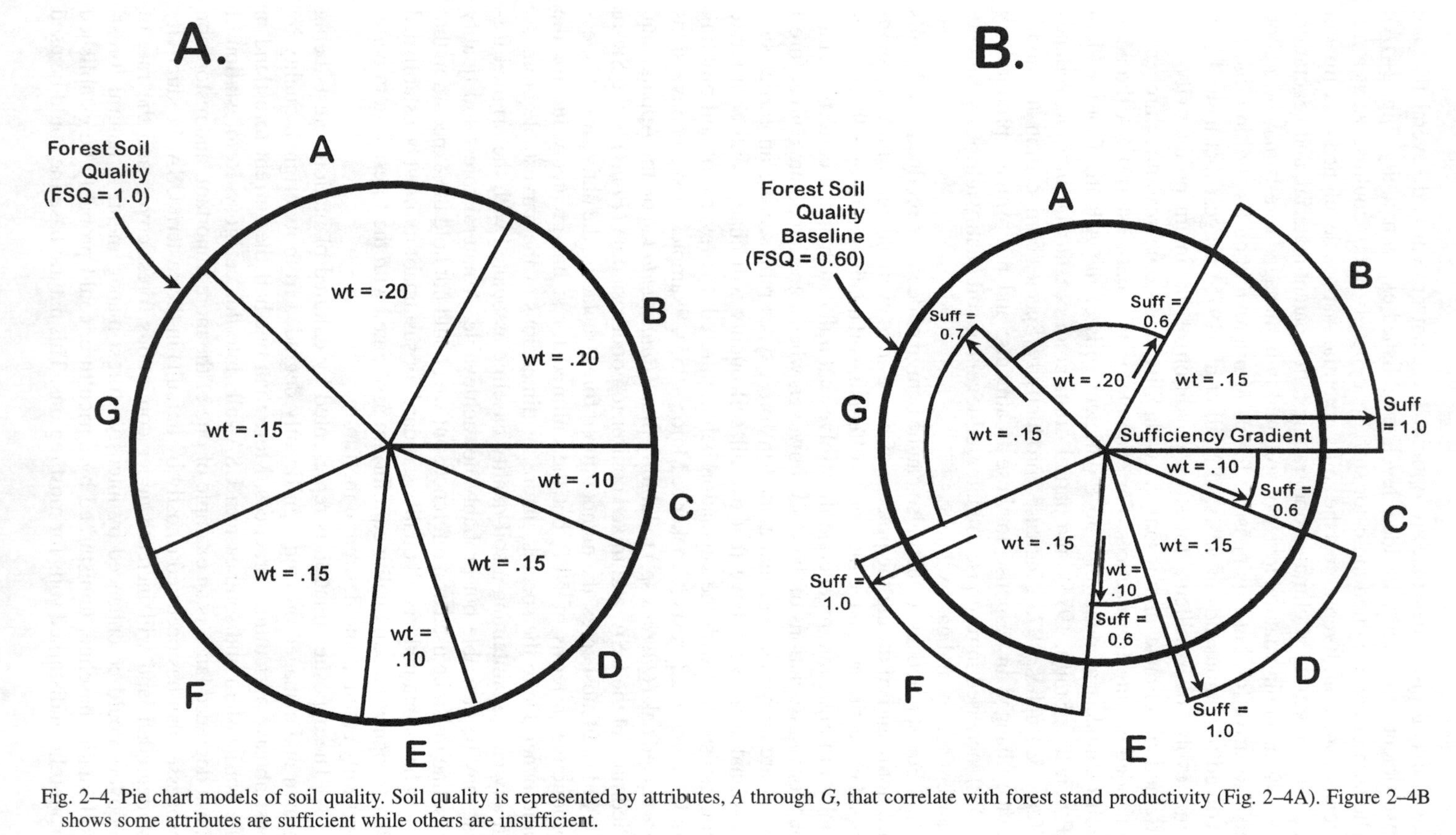

Fig. 2–4. Pie chart models of soil quality. Soil quality is represented by attributes, *A* through *G*, that correlate with forest stand productivity (Fig. 2–4A). Figure 2–4B shows some attributes are sufficient while others are insufficient.

have several properties that are optimal for plant growth and several that are insufficient compared with ideal levels established experimentally (Fig. 2–4A). The sum of the weighted indicator sufficiency levels (0.60) could serve as a reference condition. It represents the soil's potential that has developed over time in a particular setting of climate, topographic relief, parent material, and vegetation.

Maintaining natural levels of soil quality in managed forests may be a good goal to strive for, but it is possible for us to improve upon a condition that has evolved over thousands of years. O'Neill et al. (1975) suggest that natural systems evolve over millenia toward a condition of maximum productivity constrained by climate and resource availability, but some resources are easily amended by forestry practices. For example, in the southeastern USA the addition of modest amounts of P to P-deficient soils of the Atlantic Coastal Plain (Pritchett & Smith, 1974), and partial drainage of coastal and interior wetlands (Terry & Hughes, 1975), increased productivity. Species composition and dominance changed in response to these treatments, and net primary productivity (NPP) increased in response to increased levels of soil quality attributes, namely nutrient supply and gas exchange.

Forest practices' effects also can be benign or decrease productivity. If soils are not disturbed during the process of harvesting, even clearcuts for total stemwood extraction will have little or no effect on soil quality (Johnson, 1994), especially in forests that regenerate themselves naturally such as the eastern deciduous forest associations of the USA; however, when some temperate climax forest types are converted and maintained in forest types representing an earlier successional sere, we believe that overall soil quality will change and equilibrate commensurately with the vegetation being cultured; the reverse of soil building under aggrading forests (Crocker & Major, 1955). Examples of such conversions are mixed oak (*Quercus* sp.) to loblolly pine (*Pinus taeda* L.) on the Appalachian Piedmont of the USA, and mixed conifer to ponderosa pine (*Pinus*) in the Sierra Nevadas. In most cases, the monocultured forest replacing the native forest is less productive in terms of NPP, but the cultured forest has greater value for the landowner due to the species, nature, or dimensions of the harvestable biomass. In any case, maintaining a soil-quality baseline associated with the original forest may be desirable but probably not achievable. A natural forest soil quality baseline might be used as a reference, but we should not hold the expectation that it can be maintained. A soil quality standard for sustainability must be established that recognizes a soil quality equilibrium associated with the forest system being presently cultured and desired in the future.

The opposite situation is represented by cultured plantation forests being grown and managed on land significantly degraded from its original quality by past abusive agricultural practices. More than one-half the current forestland in the northeastern and southeastern USA falls into this category. Reforestation of these degraded lands is an example of one of the more important land restoration processes that has occurred (mostly by default) in the eastern USA. Productivity of degraded land will increase under continuous forest cover, and the rate of recovery could be enhanced by intensive forest management. A natural forest-soil-quality baseline, undisturbed by agriculture, could probably be established from relic undisturbed soils for most regions. This natural baseline could be used as a potential to work towards as soil quality recovers from its damaged condition in response to reforestation and forest management.

Soil-, Site-, and Management-Specific Indicators

Soil quality indicators must be soil-, site-, and management-practice-specific. Indicators chosen for monitoring soil quality on young, droughty Entisols derived from marine sand dunes would not be the same as those chosen for monitoring soil quality on older, poorly-drained Alfisols derived from marine lacustrine deposits. This is illustrated in Fig. 2–5, where a soil quality gradient is hypothesized as a function of soil fertility and soil water–air balance. As fertility is depleted (e.g., soil organic matter, nutrient availability, buffering capacity, and others) and soil water–air become imbalanced (e.g., water holding capacity, conductivity, aeration porosity, drainage, and others), soil quality decreases and at some point productivity is judged nonsustainable (Fig. 2–5). For forestland in the southeastern USA, the effects of forest practices on the quality of well-drained Entisols are primarily a function of soil fertility factors, while effects on Alfisols are mostly due to management-induced changes in soil water–air balance. Due to their collective properties, other soils would fall between these two. Because of these inherent differences, indicators must be chosen specifically for a given soil and site. The relative importance of indicators between soil groups becomes less at high levels of inherent soil quality, illustrated by the convergence of the soil groups toward the upper right in the diagram.

To illustrate the soil-specificity of indicators, several managed-forest examples are superimposed on the response surface of Fig. 2–5. Forest A depicts slash pine (*Pinus elliottii* var. *elliottii*) plantations established via mechanical site

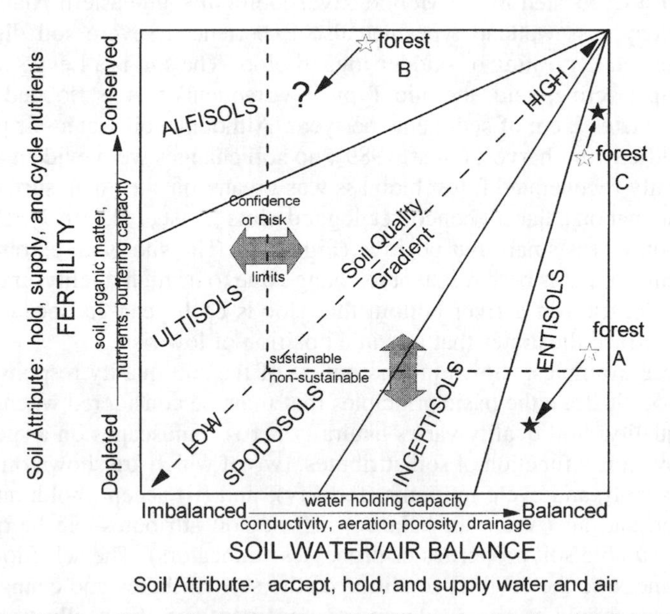

Fig. 2–5. Spatial soil quality gradient defined by soil fertility and soil water–air balance.

preparation on sites originally occupied by sand pine (*Pinus clausa* Vasey) and turkey oak (*Quercus laevis* Walt.) in the sandhills region of northwestern Florida. Fire and mechanized site preparation that removed significant amounts of organic matter and topsoil from excessively-well-drained Entisols was nonsustainable management (shown by response vector, Fig. 2–5) based on poor growth and yield of the stands (Brendemuehl, 1967). Site clearing and tillage reportedly depleted nutrient reserves on this naturally low-quality site. To help judge the sustainability of these forest operations now and in the future, appropriate indicators of soil quality for this soil, site, and management practice might include O horizon biomass, soil organic matter content, N availability, and extractable phosphorus. Indicators of fertility are more important than water–air balance, given the nature of the soil and the effect of the management practices on soil attributes.

We are currently doing sustainability research on forest B depicted in Fig. 2–5. Forest B, located on the coastal plain of South Carolina, is a 1-yr-old loblolly pine plantation planted after harvesting a 20-yr-old plantation of the same species. The forest sites are jurisdictional wetlands with highly fertile soils consisting of Argent (fine, mixed, thermic Typic Ochraqualf) and Santee (fine, mixed, thermic Typic Argiaquoll) formed on marine sediments. Wet-weather logging severely disturbed the soil and changed the water table level and rate of soil drainage. The trajectory and extent of long-term forest response are unknown, but several indicators being used to measure soil quality change are water table depth, hydraulic conductivity, aeration porosity, soil O_2 level, soil respiration, N mineralization, and redox potential. They were chosen based on their response to harvest disturbance and the dominant role they play in forest productivity on these soil and site types. Changes in these indicators are being monitored and SQI is being correlated with stand response and validated (Kelting et al., 1999).

Forest C, located in the Mobile River Delta in southeastern Alabama, is a tupelo-baldcypress wetland type that also experienced severe soil disturbance during clearcut harvesting of skidder-logged plots. The soil is a Levy series (fine, mixed, super-active, acid, thermic Typic Hydraquent) that is flooded annually and accumulates 2 cm of sediments per year. Although soil water–air properties were modified after harvest (Aust, 1989), no soil changes were evident after 7 yr, and naturally-regenerated forest biomass was greater on sites with surface disturbance than that on adjacent helicopter-logged plots (Aust et al., 1997). This short-term response was small, but positive (Fig. 2–5). The site is an extremely productive Entisol that is both resistant to change due to its high quality, and resilient due to its location in a river bottom that floods each year. Its location on the response surface illustrates that it is in a position of low risk.

These examples, shown in the context of the soil quality response surface in Fig. 2–5, illustrate the basic principles that must be considered when monitoring soil quality: Soil quality varies naturally across landscapes on a spectrum of high to low. It is a function of soil attributes, two of which are shown on the axes: (i) hold, supply and cycle nutrients (fertility), and (ii) accept, hold, and supply soil, water, and air (water–air balance). These soil attributes can be quantified using measurable soil properties or processes (indicators). The selection of indicators to measure the original condition of the soil attributes and changes due to management should be site-, soil-, and management-specific as illustrated by the location of certain soil groups on the response surface based on their characteristics. Indicator specificity becomes more important at low levels of soil quality. At some defined level, soil quality is judged sustainable (upper right area of response surface) or non-sustainable based on a soil's condition. Undisturbed forest soils are considered sustainable since they, with their associated forests, evolved into their present condition and have withstood the test of time (Kimmins, 1996). Management can change soil quality (direction and length of vector). Soils at greatest risk of damage, and soils with the greatest opportunity for improvement by management, are adjacent to the sustainable–nonsustainable line. Soils at low risk reside in the far upper right of the response surface; they are both resistant to and resilient of change and so are at low risk. Sustainability level on the response surface must be defined in terms of soil function (e.g., capacity for plant production) and be correlated with productivity within certain limits of confidence or risk (broad, shaded arrows) in order to make definitive judgments about the sustainability of forest management practices.

IMPLEMENTING A SOIL QUALITY MONITORING SYSTEM

The first step in designing a soil quality monitoring program is to qualitatively describe the attributes of a high-quality soil. Soil quality is defined based on its capacity to perform a certain function. To promote forest growth, soil should (i) allow unimpeded root growth, (ii) accept, hold, and regulate water and air to optimize delivery to plants and soil animals, (iii) store, supply, and cycle nutrients at levels and rates that are synchronized with demand, and (iv) facilitate biological activity to maintain necessary symbiotic relationships and promote nutrient cycling (Table 2–2). Implementing a soil quality monitoring program requires converting often loosely-defined qualitative descriptions of soil quality into meaningful and consistent quantitative measures, or indicators, of soil quality.

The second step is to substitute quantitative measurements for the qualitative soil quality attributes. For example, Singh et al. (1992) developed a Soil Tilth Index as a method for quantitatively measuring soil tilth. They made the transition from the qualitative to the quantitative by substituting bulk density, soil strength, aggregate uniformity, organic matter content, and plasticity index for the qualitative attributes of soil tilth. The soil tilth index was calculated using a multiplicative model made up of sufficiency relationships among crop production and each of the variables. Singh and coworkers found that their soil tilth index was significantly and positively correlated with crop yields. Thus, in viewing each attribute of soil tilth separately, they were successful in converting a qualitative description of soil tilth into a quantitative measurement of an important attribute of soil quality.

Karlen and Stott (1994) provided an excellent illustration of these steps in their example of evaluating soil quality with respect to a soil's resistance to erosion. They describe the attributes of a high-quality soil that will resist erosion as a soil that (i) freely allows water infiltration, (ii) has optimum water transfer and adsorption characteristics, (iii) is resistant to physical damage, and (iv) promotes

Table 2–3. Soil quality attributes and associated indicators.

Attribute	Weight	Indicator	Weight
Water infiltration	0.50	Infiltration rate	1.00
Water transfer and adsorption	0.10	Hydraulic conductivity	0.60
		Porosity	0.15
		Macropores	0.25
Resist physical damage	0.35	Aggregate stability	0.80
		Shear strength	0.10
		Soil texture	0.05
		Heat transfer capacity	0.05
Promote plant growth	0.05	Rooting depth	0.25
		Water relations	0.35
		Nutrient relations	0.30
		Chemical barriers	0.10

plant growth. They then developed a list of measurable soil quality indicator variables that could substitute for each attribute (Table 2–3). With the exception of water infiltration, it is necessary to use several indicators in describing each soil quality attribute. In the situations where several indicators are used, Karlen and Scott integrate them into an index of the respective attribute using an approach similar to that of Singh et al. (1992). Notice there are no direct measurements for the indicators chosen to substitute for the "promote plant growth" attribute. In this situation, measurable indicators must be substituted for the nonmeasurable indicators; for example, measurements of organic C, pH, macronutrients, and micronutrients are substituted for the nutrient relations indicator.

After defining and weighting the attributes of high soil quality, and identifying and weighting measurable indicators for each attribute, the indicators are combined into an overall index of soil quality (Fig. 2–4). By monitoring changes in the overall index through time, positive and negative management impacts on soil quality can be determined. If certain management practices are improving soil quality, then the index will increase over time. Tracking changes in the soil quality index through time under different management scenarios will allow us to identify and choose management strategies that result in improved soil quality, while avoiding those that decrease soil quality.

Soil Quality Models

Early Soil Quality Models

An early method for combining soil attributes into an overall index of soil quality is the Productivity Index (PI) model developed by Kiniry et al. (1983). The PI model is a multiplicative model that integrates field measurements taken for several soil variables into an index that relates to plant productivity. The PI model is constructed based on the assumptions that (i) there is a direct positive relationship between root growth and aboveground productivity, and (ii) any soil property that restricts root growth will result in decreased aboveground productivity. Based on the PI model, the roots will assume an ideal distribution if no soil restrictions to root growth occur with depth, and if soil restrictions do occur with depth, the roots will negatively deviate from the ideal distribution. Kiniry et al. (1983) chose five soil variables to include in their PI model: (i) available water capacity, (ii) bulk density, (iii) aeration, (iv) pH, and (v) electrical conductivity. The PI was then calculated using the model:

$$\mathrm{PI} = \sum_{i=1}^{d} (A \times B \times C \times D \times E \times \mathrm{RI})_i \qquad [1]$$

where A, B, C, D, and E are values determined from sufficiency relationships developed for each variable with respect to root growth (e.g., Fig. 2–6), and RI is a weighting factor based on the ideal root distribution. The PI is calculated by

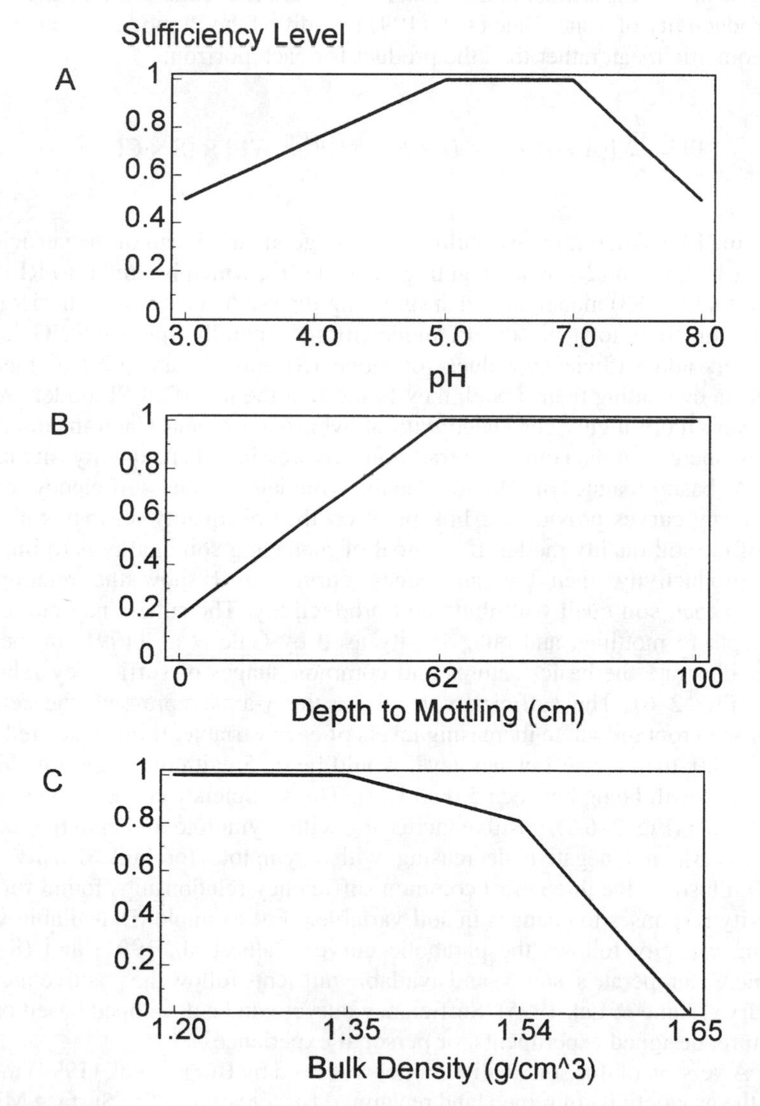

Fig. 2–6. Sufficiency curves for (A) pH, (B) depth to mottling (aeration index) and (C) bulk density (Gale et al., 1991).

summing the product of the five variables times the weighting factor with depth. The sufficiency relationships and the RI are standardized between 0 and 1, so the PI is between 0 and 1. As the PI approaches 1, the root distribution approaches the ideal root distribution, and the site productivity increases. Because the PI model is based on the ideal root distribution, it must be modified for different plant species.

One problem with the PI model is that if all the soil variables are at the same level in one horizon, then their product will be lower than their individual sufficiency values; e.g., if all five sufficiency values are 0.95, then their product would be 0.77. This artifact of the model may cause it to underestimate the potential productivity of a site. Gale et al. (1991) modified the PI model by calculating the geometric mean rather than the product for each horizon:

$$\mathrm{PI} = \sum_{i=1}^{d} [(A \times B \times C \times D \times E \times \mathrm{RI})^{1/4} \times \mathrm{WF}] \times (\mathrm{S} \times \mathrm{CL})^{1/2} \qquad [2]$$

where the PI is determined by multiplying the geometric mean of the sufficiency values (A, B, C, and D) by a weighting factor (WF), which is similar to RI in the Kiniry et al. (1983) model, and then summing the products across all horizons (i) present. In order to make the PI model more generally applicable, Gale and coworkers add sufficiency values for slope (S) and climate (CL) to their PI model. In evaluating their model, they found that the modified PI model predictions were more highly correlated with aboveground biomass and mean annual biomass increment than our more traditional expression of site quality, site index.

A basic feature of all soil quality models is the sufficiency curve. Sufficiency curves provide the link between the soil quality attributes and the goal of the soil quality model. If the goal of managing soil quality is to improve plant productivity, then the sufficiency curves must show the relationship between each soil quality attribute and productivity. The sufficiency curves for pH, depth to mottling, and bulk density used by Gale et al. (1991) in their PI model illustrate the basic features and common shapes of sufficiency relationships (Fig. 2–6). The sufficiency level on the y-axis represents the relative response in root growth to increasing levels of each variable. If the measured field pH was 4.0, then the sufficiency level would be 0.75, with the optimum soil pH for root growth being between 5.0 and 7.0. The sufficiency curves shown, parabolic for pH (Fig. 2–6A), positive increasing with asymptote for depth to mottling (Fig. 2–6B), and negative decreasing with asymptote for bulk density (Fig. 2–6C), illustrate the three most common sufficiency relationships found for productivity responses to changes in soil variables. For example (i) available water holding capacity follows the parabolic curve (Gale et al., 1991) and (ii) soil organic C (temperate systems) and available nutrients follow the positive increasing curve (Aune & Lal, 1995). Sufficiency curves can be developed based on the literature, designed experiments, or personal experience.

A version of this soil quality model was used by Burger et al. (1994) to estimate the productivity of mined land reclaimed for forestland. The Surface Mining Control and Reclamation Act of 1977 requires surface miners to restore mined sites to their original use and to their original level of productivity. In the Central Appalachian region, mixed oak forests are removed and are commonly replaced with white pines (*Pinus strobus* L.), a mid-successional native species. Topsoil replacement is normally required, but because of the steep terrain and shallow soils of the region, miners obtain a waiver that allows them to apply a topsoil substitute made up of a mixture of recoverable soil and overburden materials. About 75 to 100% of the surface 1 m commonly consists of C_r-horizon material.

When forage or cropland is restored, the productivity requirement is met by achieving sustained forage or crop yields (two successive years) comparable to those produced on undisturbed soils. Comparable bioassays or comparative assessments of 2- to 5-yr-old trees are not useful productivity standards for trees. Essentially, no productivity standard was available for trees. Therefore, an attempt was made to develop a soil-based standard for mined land designated for reforestation.

Preliminary studies on mine soils as media for plant growth in the region suggested that the major growth limiting factors were soil density due to compaction (Daniels & Amos, 1984), P deficiency due to high soil P-fixing capacities (Roberts et al., 1988), and high levels of soluble salts (Torbert et al., 1989). For each of these soil properties, a literature-derived sufficiency function was developed quantifying the influence of each property on tree growth. Growth values were normalized between 0.0 and 1.0, where 0.0 reflected a condition that completely inhibited root growth, and 1.0 reflected a condition that was ideal for root growth. Phosphorus sufficiency was based on $NaHCO_3^-$-extractable P (Olsen & Sommers, 1982) and calculated according to the following equations:

$$\text{If P} > 9 \text{ mg kg}^{-1}\text{: P sufficiency} = 1.0;$$

$$\text{If P} < 9 \text{ mg kg}^{-1}\text{: P sufficiency} = 0.29 + 0.32ln(\text{P})$$

Andrews (1992) developed a relationship between white pine growth and electrical conductivity (EC). Based on his work, the following EC sufficiency equations were developed:

$$\text{EC} <0.5\text{, EC Sufficiency} = 1.0;$$

$$\text{EC} >0.5\text{, EC Sufficiency} = 1.26 - 0.53(\text{EC})$$

pH sufficiency was based on relationships developed by Gale et al. (1991):

$$\text{pH} <5.0\text{, pH sufficiency} = -0.25 + 0.25 \times \text{pH};$$

$$\text{pH 5.0 to 6.0, pH sufficiency} = 1.0;$$

$$\text{pH} >6.0\text{, pH sufficiency} = 2.5 - 0.25(\text{pH})$$

The weighting factor (WF) is a measure of soil depth or rooting volume sufficiency. It is an estimate of the proportion of roots within a given soil layer if all conditions are optimum. A rooting distribution curve developed by Gale and Grigal (1987) for mid-tolerant conifers was used:

$$Y = 1 - 0.96^{d},$$

where Y is the cumulative root fraction from the surface to soil depth d in centimeters. A PI model of the form developed by Gale and Grigal (1991) was used to combine sufficiency estimates to provide an index of productivity for restored mine soils:

$$PI = (P \times EC \times pH)^{1/3} \times WF$$

As an initial check on the performance of each of the selected soil quality indicators, the heights of 10-yr-old white pines growing on 36 different mine soils were measured along with the soil quality indicators. PI was calculated for each site and correlated with corresponding tree height. The relationship was highly significant ($P < 0.02$).

The PI was then calibrated with average white pine site index for the region in order to use it as a productivity standard. Average site index for white pine at age 50 in the region is 80, with SI 100 representing the best undisturbed sites. Since SI 100 would be about the maximum productivity achievable for mine soils, it should correspond to a PI of 1.0. Assuming a linear relationship between SI and PI, the average SI of 80 for the region would correspond to a PI of 0.8. An average site index was calculated for the 10-yr-old white pines in the study sites using Beck's (1971) equation for polymorphic SI curves, and a PI was calculated from the average mine soil chemistry and depth corresponding to SI 80:

$$pH = 4.5\text{: pH sufficiency} = -0.25 + 0.25(4.5) = 0.89$$

$$EC = 0.89 \text{ dS m}^{-1}\text{: EC sufficiency} = 1.26 = 0.53(0.89) = 0.79$$

$$P = 10.2 \text{ mg kg}^{-1}\text{: P sufficiency} = 1.00 \text{ since } P > 9 \text{ mg kg}^{-1}$$

$$\text{depth} = 72 \text{ cm: WF} = 1 - 0.96^{72} = 0.94$$

$$PI = (0.89 \times 0.79 \times 1.0)^{1/3} \times 0.93 = 0.82$$

Based on this study, SI 80 corresponded to a PI of 0.82; thus, a reclamation productivity standard of PI = 0.8 is a reasonable goal to strive for in order to return the mined land of the region to an approximation of its original level of productivity as required by law.

This study demonstrates a first attempt at estimating productivity based on soil quality indicators in a situation where bioassay or comparative productivity estimators were not available. Obviously, more work could be done to select and include additional indicators of mine soil quality that would better reflect potential for tree growth; however, in any environment, there is a myriad of soil properties and processes controlling production over time. The objective is not to attempt to quantify every factor (which is not achievable), but rather to adequately index productivity with a minimum data set of key soil indicators.

Models with Indicator Weighting and Indicator Constructs

Another method for determining an overall index of soil quality was developed by Karlen and Stott (1994). Soil quality (Q) is determined using an additive model:

$$Q = q_1\,(\text{wt}) + q_2\,(\text{wt}) \ldots + q_k\,(\text{wt}) \qquad [3]$$

where the q_k variables represent sufficiency values for different soil quality attributes, and the wt are relative weights applied to each attribute. The relative weights represent the importance of each attribute in determining overall soil quality. In applying this model to the resistance to erosion soil quality attributes (Table 2–3), the model would take the form:

$$Q = q_{wi}\,(0.50) + q_{wt}\,(0.10) + q_{rd}\,(0.35) + q_{pg}\,(0.05) \qquad [4]$$

where q_{wi}, q_{wt}, q_{rd}, and q_{pg} are sufficiencies for water infiltration, water transfer and adsorption, resistance to physical damage, and promoting plant growth, respectively, and the numbers are their relative weights (Table 2–3). The performance of this model has not yet been tested using field data.

Additive models (Eq. [3] and [4] and Fig. 2–5) have been criticized because they may not allow for interaction among the variables in the model. Gale et al. (1991) argue that multiplicative models (Eq. [1] and [2]) are better than additive models because multiplicative models account for the possibility of interaction among the model components, and therefore more closely follow basic biological principles.

There is no question that many soil properties and processes interact at some level, and soil quality models should consider interactions where applicable; however, rather than arbitrarily building either a multiplicative or an additive model of soil quality, the extent of interaction among the variables in a soil quality model should be controlled based on current understanding of how the various soil components interact. For example, if we know *a priori* that two or more components interact, as in the PI model (Eq. [1] and [2]), then the components should be combined into a single expression using a multiplicative model. These single expressions derived from multiple components are examples of soil quality indicator constructs, or pedotransfer functions (Bouma, 1989).

The Soil Tilth Index (Singh et al., 1992) previously discussed is a good example of an indicator construct or pedotransfer function that combines five interacting soil variables into a single expression. Karlen and Stott's (1994) soil quality model (Eq. [3] and [4]) is actually a hybrid located somewhere between a multiplicative and an additive model. Three of the indicators in their model are constructs of several soil properties, calculated based on individual multiplicative models. For example, the water transfer and adsorption attribute (Table 2–3, Eq. [4]) is calculated by multiplying sufficiency values for hydraulic conductivity, porosity, and macroporosity together. In addition to the individual multiplicative models representing some attributes, by assigning relative weights to each attribute (Eq. [3] and [4)], Karlen and Stott are controlling the importance of each attribute in their soil quality model.

Forest Soil Quality Model

The PI model (Eq. [2]) and Karlen and Stott's (1994) soil quality model (Eq. [3]) have desirable elements that should be incorporated into a forest soil quality model. Timber harvesting and site preparation can result in soil disturbances (e.g., compaction, rutting, and churning), which may limit both root growth and overall tree growth (Greacen & Sands, 1980; Kozlowski et al., 1991). Therefore, the concept of an ideal root distribution, which is the basis of the PI model, is important to forest soil quality because our management practices can affect tree root distributions. The soil quality attributes listed in Table 2–2 for forest productivity are necessary for maintaining high levels of productivity; however, the relative importance of each attribute is not the same for all forests growing under all conditions; i.e., management intensity, soil type, climate, and topographic features all play roles in determining the relative importance of each soil quality attribute. For example, the promote root growth attribute would be more important to soil quality for soils with inherently high soil strength (Ultisols in the Piedmont Region), vs. soils with inherently low soil strength (Alfisols in the Coastal Plain Region). Because of such differences in the importance of soil attributes, relative weights need to be assigned to the soil attributes, which was a concept introduced in the soil quality model (Eq. [3]). The forest soil quality (FSQ) model we propose incorporates the features of the PI and soil quality models (Eq. [2] and [3]):

$$\mathrm{FSQ} = \sum_{i=1}^{d} [(\mathrm{RG} \times \mathrm{wt}_{\mathrm{RG}}) + (\mathrm{WS} \times \mathrm{wt}_{\mathrm{WS}}) + (\mathrm{NS} \times \mathrm{wt}_{\mathrm{NS}}) + (\mathrm{GE} \times \mathrm{wt}_{\mathrm{GR}}) + (\mathrm{BA} \times \mathrm{wt}_{\mathrm{BA}})]\mathrm{WF}_{\mathrm{d}} \qquad [5]$$

where RG, WS, NS, GE, and BA are the sufficiency levels for promoting root growth, water supply, nutrient supply, gas exchange, and biological activity, respectively, and the wts are the relative weights of each attribute. For soil quality attributes that cannot be measured, pedotransfer functions would be used in the model to substitute for the nonmeasurable attributes. Forest soil quality, FSQ, is determined by multiplying the sum of the weighted sufficiency values by a weighting factor, WF, the relative volume of each soil horizon, and then summing the products across all horizons, *i*, present. The FSQ values will range between 0 and 1, with 1 being ideal soil quality for tree productivity.

Within genetic constraints, forest productivity depends on the integration of multiple factors that together define site quality. These multiple factors include but are not limited to topographic features (e.g., slope and aspect), climate, and soil quality. Because soil quality is a subset of site quality, soil quality can be used by itself to index forest productivity only when site factors are negligible. In relatively flat terrain of the Atlantic and Gulf Coastal Plains, soil is the dominant abiotic factor affecting forest productivity. In steep mountainous areas, topographic features may dominate. In order to relate forest soil quality to forest productivity, forest soil quality must be used as one component in a forest productivity index model (FPI),

$$\mathrm{FPI} = \mathrm{FSQ} \times (S \times A \times C)^{1/3} \qquad [6]$$

where the forest productivity index (FPI) is determined by multiplying the forest soil quality index (FSQ) times the product of the sufficiency levels for slope (S), aspect (*A*), and climate (*C*). The geometric mean of the product of *S*, *A*, and *C* is used as recommended by Gale et al. (1991). The model can be easily modified to include any other additional site quality factors.

Monitoring Forest Soil Quality

A forest soil quality monitoring system should function at several levels to (i) assess the status of soil quality, (ii) monitor the effects of management practices, air pollution, and other human-caused processes on soil quality, and (iii) adapt and improve management approaches with new information (Fig. 2–7). After the indicator variables are chosen and the appropriate weights are applied, the status of soil quality is determined at the pedon level. Pedon level soil quality is then extrapolated to the stand or landscape level, where areas of low soil quality are identified and management recommendations are made to improve soil quality. Soil quality is then monitored through time in order to (i) monitor the effectiveness of management recommendations, (ii) determine if the goals are being achieved, and (iii) validate the sufficiency curves and the soil quality model. Adaptations to management strategies and the soil quality model are made based on the information obtained through monitoring.

The forest soil quality values predicted using the FSQ model represent only one of three important components of forest soil quality, with the other two com-

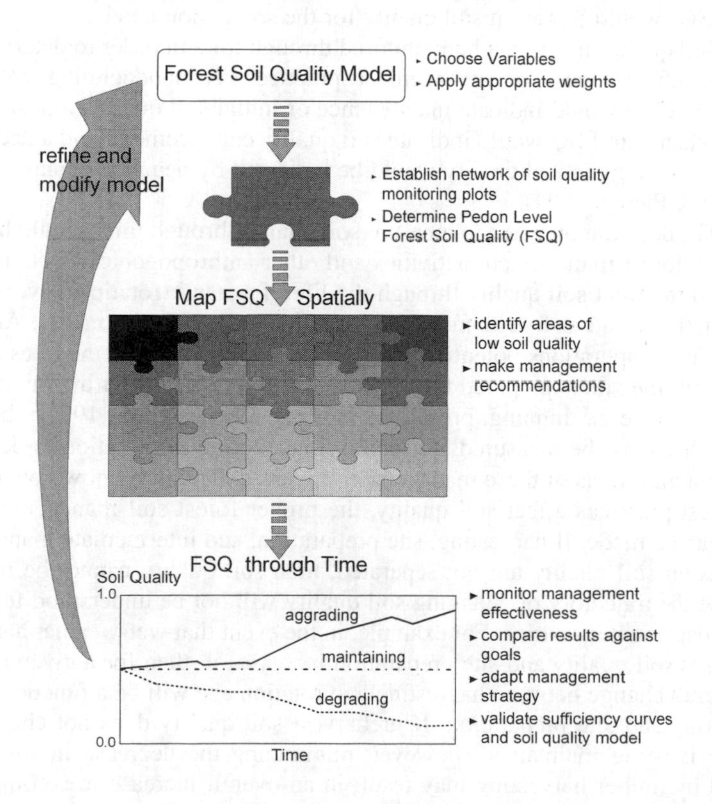

ponents being space and time (Fig. 2–7). The FSQ model predicts pedon level soil quality; however, because forest management activities are not conducted at the pedon level, but rather at the stand and ultimately the landscape level, soil quality must be assessed spatially in order to extrapolate pedon level soil quality to landscape level soil quality. In addition, the forest soil quality attributes are not static variables, but rather they change through time in response to forest management activities and natural processes.

In order to extrapolate pedon level soil quality to landscape level soil quality, a systematic grid of soil quality monitoring plots would have to be established across the landscape. The number of plots would depend on the desired management resolution (e.g., 1 ha, 5 ha), with the number of plots increasing as the required level of management resolution increased. Resolution would be a function of forest management type (Fig. 2–1) among other things. Soil quality would be determined for each plot (pedon), and geostatistical techniques (e.g., kriging) would be used to produce forest soil quality maps (Davis, 1986; Smith et al., 1993; Halvorson et al., 1996).

We suggest establishing a network of permanent plots across management units for monitoring FSQ. This network would be analogous to the U.S. Forest Service's Continuous Forest Inventory (CFI) plots used to monitor the status of our nations timber supply. The soil quality indicators would be measured at each plot, and from these data individual plot FSQs would be calculated. These plot-level FSQs would represent soil quality for the soil pedon level.

Soil quality must then be monitored through time in order to determine the ultimate effectiveness of management practices. A nondeclining pattern of change in FSQ would indicate maintenance of initial soil quality, a positive pattern of change in FSQ would indicate soil quality enhancement, and a decreasing pattern of change in soil quality would be indicated by negative changes in FSQ (Larson & Pierce, 1994).

The decision on when to measure soil quality through time should be based on when forest management activities and other anthropogenic effects have the potential to affect soil quality through the life of a stand (rotation), over successive rotations, and as forests managed for nontimber products mature. At a minimum, forest operations potentially affect soil quality when (i) the trees are harvested, (ii) the site is prepared for the next rotation, and (iii) during intermediate management (e.g., thinning, prescribed burning; Powers et al., 1990); therefore, soil quality must be measured at specific times during the rotation to determine the different effects of these management practices. Once we know how individual forest practices affect soil quality, the proper forest soil management decisions can be made. If harvesting, site preparation, and intermediate management impacts on soil quality are not separated, then soil quality cannot be managed because the trajectory of changing soil quality will not be understood for different management scenarios. For example, in the event that wet-weather harvesting decreases soil quality and site preparation increases it, then for a given soil type the overall change between harvesting and rotation age will be a function of both harvesting and site preparation. If preharvest soil quality does not change, soil quality is being maintained; however, minimizing the decrease in soil quality caused by timber harvesting may result in an overall increase in soil quality. If timber harvesting and site preparation effects on soil quality are not separated, then neither would the positive effects of site preparation, nor the negative effects of harvesting, be realized.

Making Forest Soil Management Decisions

Constructing soil quality maps based on the pedon level FSQ values shows the status of soil quality, but it would be difficult to make forest soil management decisions based on the FSQ values shown on the map. For example, an area of the soil quality map shows an average FSQ of 0.6, indicating the soil in this area is at only 60% of its potential for tree production. The question is, what can the forest manager do to increase soil quality beyond an FSQ of 0.6? Halvorson et al. (1996) developed a technique for making go, no-go soil management decisions based on predefined standards for acceptable and unacceptable soil quality. With their technique, managers can identify areas of unacceptable soil quality, determine which soil attributes are causing soil quality to be unacceptable, and make recommendations to improve the unacceptable attributes.

With their method, the sufficiency value for each soil quality attribute or pedotransfer function is converted from a continuous to a binary (i.e., 0 or 1) variable based on critical thresholds determined for each variable. For example, the critical threshold for each variable may be 80% of maximum sufficiency (or 0.80). Based on the critical thresholds, the data is converted into a binary dataset at the pedon level by coding each variable as 1, if the variable is below the critical threshold (i.e., unacceptable), and 0, if the variable is above the threshold level (i.e., acceptable). Then the pedon level binary data are combined into a multivariable indicator transform (MVIT), by coding soil quality at the pedon level as 1 (unacceptable), if one or more soil quality attributes are unacceptable, or 0, if all soil quality attributes are acceptable. Kriging is then used to project the pedon level MVIT's to unsampled locations; the kriged values are weighted means between 0 and 1 and represent the probability of an area having unacceptable soil quality.

After probability maps of unacceptable soil quality are constructed based on the kriged estimates, the soil quality attribute(s) responsible for the unacceptable soil quality can be identified by constructing probability maps for each attribute using the binary data set developed for the attributes. The areas of unacceptable soil quality on the attribute maps will correspond to the areas of unacceptable soil quality on the overall soil quality map constructed using the MVITs. Corrective practices can then be applied to improve soil quality in the unacceptable areas. For example, if soil quality is identified as unacceptable in an area because the promote root growth attribute is below the threshold level, the manager may choose to harrow or subsoil the site to decrease soil strength.

Validating and Refining Soil Quality Monitoring Systems

The usefulness of the FSQ predictions will be a function of (i) the appropriateness of indicators chosen to represent each attribute, (ii) the accuracy of the individual sufficiency curves, and (iii) the design of the FSQ monitoring system.

The criteria used to select appropriate indicators were previously discussed, and one criterion was that the indicators must have known relationships with productivity. It is necessary to know the relationship between the indicators and productivity so that sufficiency curves can be constructed for each indicator. First approximations of sufficiency curves can be made based on the literature; however, sufficiency curves can be validated and improved by measuring tree growth at the same time the soil quality indicators are measured. By pairing tree growth data with the indicator measurements, regression techniques can be used to fit better sufficiency curves as well as increase our understanding of the relationships between the soil quality attributes and tree productivity.

In order to (i) select the right indicators, (ii) apply the appropriate weights, and (iii) make the correct management decisions, definitive field experiments would be necessary for validating relationships among the indicators, management strategies, and forest productivity. These field experiments would be located on representative soil types, and would test several alternative management scenarios. Results from the field experiments would increase the level of confidence in the soil quality monitoring system by supplementing the management-level data with more scientifically controlled data.

An example of a field experiment designed to determine (i) the appropriateness of several soil quality indicators, (ii) the relative importance of each indicator, and (iii) the effectiveness of different site preparation methods in mitigating soil quality is being conducted in loblolly pine plantations located on the Lower Coastal Plain in South Carolina, on soils that are often disturbed during timber harvesting. The overall hypotheses being tested are that (i) rutting and churning may result in long-term declines in timber productivity if conventional site preparation techniques are not successful in mitigating soil disturbance, and (ii) organic matter displacement during stem removal may further exacerbate the soil disturbance problem by decreasing nutrient availability and soil tilth recovery.

In 1991 three study blocks were subdivided into six 3-ha plots, and two operational harvesting treatments were randomly installed on five plots per block: (i) two dry-soil harvests, and (ii) three wet-soil harvests. Three levels of site preparation: (i) none, (ii) conventional bedding, and (iii) mole plowing then bedding, were randomly installed on the wet harvested plots, and two levels of site preparation were randomly assigned to the dry harvested plots: (i) none, and (ii) conventional bedding. The additional plot was left unharvested to serve as a control. Loblolly pine seedlings were hand-planted on the plots in February of 1996.

The timber harvesting treatments created gradients in organic matter and soil disturbance. It was hypothesized that these gradients would alter soil quality, resulting in long-term changes in loblolly pine productivity. To test this hypothesis, a 5 by 5 organic matter–soil disturbance matrix was defined (Fig. 2–8), and mapped spatially on each wet and dry plot following harvest. Several soil quality indicators are being measured in three replications of each cell in the 5 by 5 matrix at locations chosen from the organic matter–soil disturbance maps (Fig. 2–9). Each indicator was chosen from the list in Table 2–2.

The indicators will be integrated into initial forest soil quality (FSQ) estimates using literature-based sufficiency curves. Functional relationships between

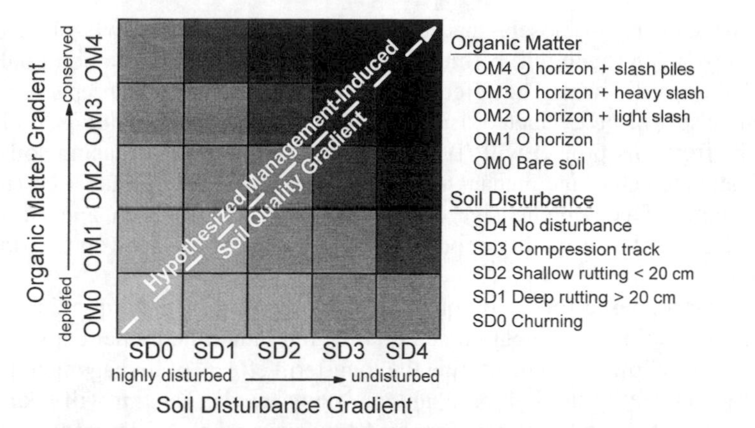

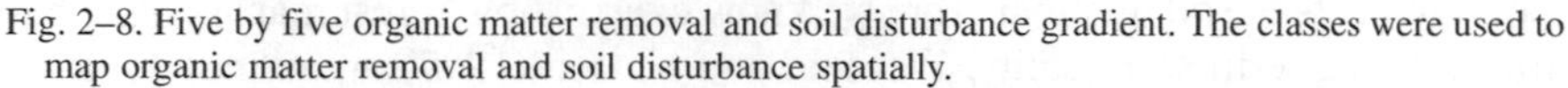

Fig. 2–8. Five by five organic matter removal and soil disturbance gradient. The classes were used to map organic matter removal and soil disturbance spatially.

FSQ, organic matter, and soil disturbance will be explored for each level of site treatment. Spatial soil quality maps will be constructed and used in combination with process modeling to predict loblolly pine productivity as a function of soil quality. The appropriateness of each soil quality indicator and the FSQ estimates will be evaluated in the short-term by comparing the sufficiency predictions based on the literature-derived sufficiency curves with the productivity of loblol-

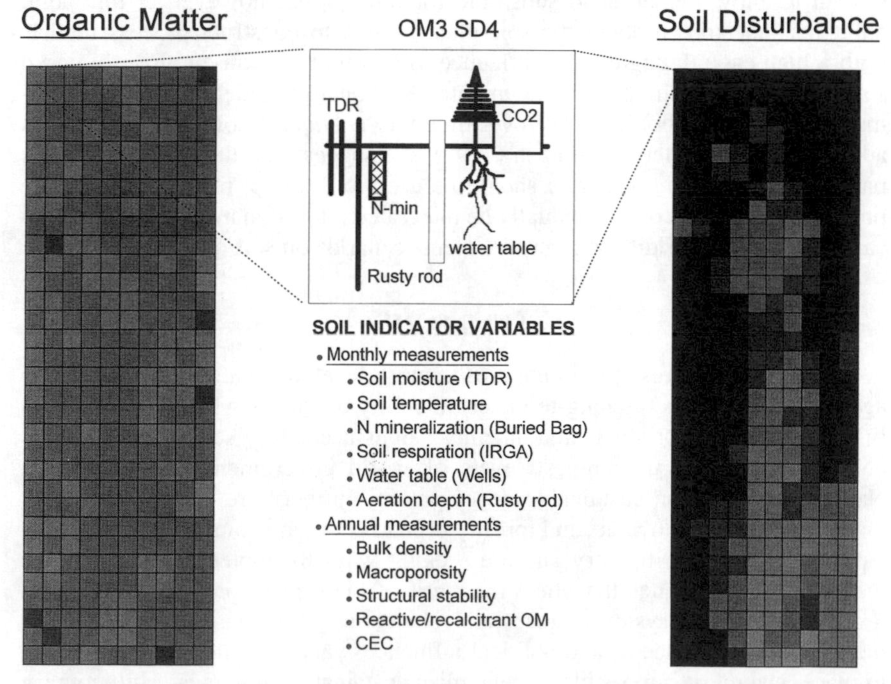

Fig. 2–9. Spatial maps of organic matter redistribution and soil disturbance following a wet site timber harvest. Soil quality indicators are being measured at representative points for each level of organic matter and soil disturbance.

ly pine growing in closely-spaced bioassay plots. As the newly-planted stand develops, tree growth will be monitored to verify and refine our soil quality and modeling predictions. The effectiveness of the different site preparation treatments also will be evaluated using the same approach. It is anticipated that the results from this project will (i) illustrate the effectiveness of using soil quality indicators to determine management effects on long-term productivity, (ii) show the benefits of a comprehensive soil quality monitoring program, and (iii) provide a useful forest soil quality model for some soils of the Lower Coastal Plain region.

Our proposed forest soil quality model (Eq. [5]) does not predict the long-term effects of management on soil quality, but rather the model combined with monitoring allows us to determine the long-term effects of management by monitoring soil quality through time and analyzing trends. Thus, it will take several measurements of soil quality before we know exactly how forest management has affected long-term soil quality. Wagenet and Hutson (1997) argue that dynamic soil quality models, rather than static models (our model), must be used to assess soil quality. Dynamic models would be composed of process models that would represent the components of soil quality. The process models would simulate the effects of management on soil quality, and as such would be advantageous over a static model because the effects of different management scenarios on soil quality could be predicted immediately from a one-time collection of field data. The dynamic soil quality model assumes that enough is known about the soil–plant system to allow modeling to substitute for monitoring; however, at this point enough is not known about the soil–plant system to construct process models with a high enough degree of confidence to substitute for an empirically-based soil quality monitoring program. Our forest soil quality model is a first approximation model that should evolve over time into a dynamic soil quality model as advocated by Wagenet and Hutson (1997). We suggest that the data gathered as part of the monitoring program should be used to develop, refine, and validate process models that could eventually be integrated into a dynamic model for forecasting the effects of different management scenarios on soil quality.

SUMMARY

The goal of forestry communities in most developed nations is forest management that ensures the long-term sustainability of forests while simultaneously providing forest products and amenity values needed by society. The goal is being advocated at the highest levels of world government through United Nations initiatives on sustainable development. Furthermore, at the local level many individual landowners and forestry organizations are committed to the principles of sustainable forestry and are seeking ways to implement and monitor management techniques that meet the spirit of these principles. Forest ecosystems, management systems, goals and objectives of management, economic incentives, societal needs, and political influences vary tremendously from place to place, making it impossible to generalize sustainable strategies. Multicountry initiatives such as the Montreal and Helsinki Processes have resulted in national-level criteria and indicators for defining and measuring sustainable forestry, but actually achieving sustainability requires the application and monitoring of forest practices at the forest site level.

Soils play an important biological role related to forest function, productivity, and protection; they must be conserved and their quality maintained in order to meet sustainability goals. Traditionally, forest soil quality and health have not been considered separately from forest quality and health. Like other forest productivity determinants such as climate, topography, and other site factors, soil has been lumped as a component of forest site and required no special consideration in management. This changed when natural forests and other land types were converted to forest plantations and tree crops using techniques that greatly manipulated soil properties and processes. Forest management now includes soil management with the aim of maintaining its quality and increasing it when possible.

Forest soil quality and forest productivity are inseparably linked and they influence each other in many ways; however, soil quality must be monitored separately from forest productivity. The most compelling argument for using changes in soil attributes or changes in indicators of soil quality to measure sustainable forest productivity is that the alternative, a measure of forest response to a given practice, is confounded by too many other factors. Measures of growth rate or biomass increment from one rotation or forest cycle to the next is a function of genotype, weather cycles, pest and disease effects as well as cultural practices. There is no way to definitively partition response among the many factors that influence forest growth in order to make a judgment about the sustainability of a given practice. Furthermore, increments of forest production require such long periods that comparisons of management practices' effects on production simply are not timely enough. Since direct measurements of management effects on long-term forest productivity are untenable, a mechanism for using short-term indicators of sustainable productivity must be sought.

Soil quality is defined as the capacity of a soil to perform a given function; its capacity is a function of its properties, the total of which determines its quality. In its undisturbed condition, it has a natural quality, and through management this quality can be enhanced or degraded. Various systems for monitoring soil quality have been proposed, primarily by agricultural soil scientists. A forest soil quality monitoring system can be devised by incorporating many principles and aspects of these agricultural-based systems. We propose a system that defines soil quality based on its ability to promote forest productivity; capture, store, and cycle carbon; capture, store, and cycle water; and bioremediate organic wastes. Development of a forest soil quality monitoring system requires a conceptual understanding of soil properties and processes, how they function together, and how they influence forest productivity. Management-induced changes in soil determinants of forest productivity must be measurable and must correlate with productivity. The sum of the changes must be integrated for a useful index of soil quality that would provide the ability to compare management effects on soil quality against an established reference condition specific for a given site type. It would allow a judgment of whether or not management practices were sustainable, and provide information for developing alternatives for a process of adaptive management.

Forest system monitoring, which amounts to interpreting the scientific literature and applying it in ways to make judgments about the efficacy of forest practices on forest productivity and soil quality, has not been a glamorous activity for forest soil scientists compared to doing original research on timely topics. However, given the potential payoff in terms of betterment of the forest condition, a "scholarship of application" (Boyer, 1990) should be encouraged to help define, describe, measure, and monitor sustainable forestry at all institutional levels for the sake of maintaining the ecological integrity of forests, the viability of forestry businesses, and the social and political acceptance of forest management. This is a role to be played by all forest soil scientists and practitioners.

REFERENCES

American Forest and Paper Association. 1995. Sustainable forestry initiative. AFPA, Washington, DC.

Andrews, J.A. 1992. Soil productivity model to assess forest site quality on reclaimed surface mines. M.S. thesis. Virginia Polytechnic Inst. and State Univ., Blacksburg.

Aplet, G.H., N. Johnson, J.T. Olson, and V.A. Sample. 1993. Prospects for a sustainable future. p. 309–314. *In* G.H. Aplet et al. (ed) Defining sustainable forestry. Island Press, Washington, DC.

Aune, J.B., and R. Lal. 1995. The tropical soil productivity calculator: A model for assessing effects of soil management on productivity. p. 499–520. *In* R. Lal and B.A. Stewart (ed) Soil management: Experimental basis for sustainability and environmental quality. Adv. Soil Sci. Lewis Publ., London.

Aust, W.M. 1989. Abiotic functional changes of a water tupelo-baldcypress wetland following disturbance by harvesting. Ph.D. diss. North Carolina State Univ., Raleigh.

Aust, W.M., S.H. Schoenholtz, T.W. Zaebst, and B.A. Szabo. 1997. Recovery status of a tupelo-cypress wetland seven years after disturbance: Silvicultural implications. Forest Ecol. Manage. 90:161–169.

Ballard, R., and S.P. Gessel (ed.). 1983. IUFRO Symp. Forest site and continuous productivity. USDA For. Serv. Gen. Tech. Rep. PNW-163. U.S. Gov. Print. Office, Washington, DC.

Beck, D.E. 1971. Polymorphic site index curves for white pine in the southern Appalachians. USDA For. Serv. Res. Pap. SE-80. U.S. Gov. Print. Office, Washington, DC.

Bouma, J. 1989. Using soil survey data for quantitative land evaluations. Adv. Soil Sci. 9:177–213.

Boyer, E.L. 1990. Scholarship reconsidered: Priorities of the professoriate. Spec. Rep. The Carnegie Foundation for the Advancement of Teaching, Princeton, NJ.

Brendemuehl, R.H. 1967. Loss of topsoil slows slash pine growth in Florida sandhills. USDA For. Serv. So. For. Exp. Stn. Res. Note SO-53. U.S. Gov. Print. Office, Washington, DC.

Burger, J.A. 1994. Cumulative effects of silvicultural technology on sustained forest productivity. p. 59–70. *In* M.K. Mahendrappa et al. (ed.) Canadian For. Serv. Maritimes Region Inf. Rep. M-X-191E.Micromedia, Hull, QU.

Burger, J.A. 1996. Limitations of bioassays for monitoring forest soil productivity: Rationale and example. Soil Sci. Soc. Am. J. 60:1674–1678.

Burger, J.A. 1997. Conceptual framework for monitoring the impacts of intensive forest management on sustainable forestry. p. 147–156. *In* O. Hakkila et al. (ed.) Forest management for bioenergy. Res. Pap. 640. Finnish Forest Res. Inst., Helsinki.

Burger, J.A., J.E. Johnson, J.A. Andrews, and J.L. Torbert. 1994. Measuring mine soil productivity for forests. p. 48–56. *In* Int. Land Reclamation and Mine Drainage Conf. Vol. 3. Reclamation and revegetation. USDOI Bureau of Mines Spec. Publ. SP 06C-94. U.S. Gov. Print. Office, Washington, DC.

Carmean, W.H. 1975. Forest site quality evaluation in the United States. Adv. Agron. 27:209–269.

Crocker, R.L., and J. Major. 1955. Soil development in relation to vegetation and surface age of Glacier Bay, Alaska. J. Ecol. 43:427–448.

de Silva, A.P., B.D. Kay, and E. Perfect. 1994. Characterization of the least limiting water range of soils. Soil Sci. Soc. Am. J. 58:1775–1781.

Daniels, W.L., and D.F. Amos. 1984. Generating productive topsoil substitutes from hard rock overburden in the southern Appalachians. Environ. Geochem. Health. 7:8–15.

Davis, J.C. 1986. Statistics and data analysis in geology. 2nd ed. John Wiley & Sons, New York.

Doran, J.W., D.C. Coleman, D.F. Bezdicek, and B.A. Stewart. (ed.) 1994. Defining soil quality for a sustainable environment. SSSA Spec. Publ. 35. ASA and SSSA, Madison, WI.

Doran, J.W., M. Sarrantonio, and M.A. Liebig. 1996. Soil health and sustainability. Adv. Agron. 56:1–55.

Dyck, W.J., and M.F. Skinner. 1990. Potential for productivity decline in New Zealand radiata pine forests. p. 318–332. *In* S.P. Gessel et al. (ed.) Sustained productivity of forest soils. Proc., 7th N. Am. Forest Soils Conf. Faculty of Forestry Publ., Vancouver, BC. 24–28 July 1988. Faculty of Forestry, Univ. of British Columbia, Vancouver.

Dyck, W.J., D.W. Cole, and N.B. Comerford. 1994. Impacts of forest harvesting on long-term site productivity. Chapman & Hall, New York.

Ford, D.E. 1983. What do we need to know about forest productivity and how can we measure it? p. 1–12. *In* R. Ballard and S.P. Gessel (ed.) Proc. IUFRO Symp. on Forest Site and Continuous Productivity. USDA For. Serv. Gen. Tech. Rep. PNW-163. USDA For. Serv. Pacific Northwest For. and Range Exp. Stn., Portland, OR.

Franklin, J.F. 1993. The fundamentals of ecosystem management with applications in the Pacific Northwest. p. 127–144. *In* G.H. Aplet et al. (ed.) Defining sustainable forestry. Island Press, Washington, DC.

Gale, M.R., and D.F. Grigal. 1987. Vertical root distributions of northern tree species in relation to successional status. Can. J. For. Res. 17:829–834.

Gale, M.R., D.F. Grigal, and R.B. Harding. 1991. Soil productivity index: Predictions of site quality for white spruce plantations. Soil Sci. Soc. Am. J. 55:1701–1708.

Gessel, S.P., D.S. Lacate, G.F. Weetman, and R.F. Powers (ed.). 1990. Sustained productivity of forest soils. *In* Proc. 7th North Am. Forest Soils Conf. Faculty of Forestry Publ., Vancouver, BC. 24–28 July 1988. Faculty of Forestry, Univ. of British Columbia, Vancouver.

Greacen, E.L., and R. Sands. 1980. Compaction of forest soils. A review. Aust. J. Soil Res. 18:163–189.

Greenland, D.J., and I. Szabolcs (ed.). 1994. Soil resilience and sustainable land use. CAB Int., Oxon, England.

Halvorson, J.J., J.L. Smith, and R.I. Papendick. 1996. Integration of multiple soil parameters to evaluate soil quality: A field example. Biol. Fert. Soils 21:207–214.

Helsinki Process. 1994. Helsinki process. *In* Proc. of the ministerial conferences and expert meetings. Liaison Office of the Ministerial Conference on the Protection of Forests in Europe, Helsinki, Finland. 16–17 June 1993. FIN-00171. Office of the Ministerial Conf. on the Protection of Forests in Europe, Helsinki.

Johnson, D.W. 1994. Reasons for concern over impacts of harvesting. p. 1–12. *In* W.J. Dyck et al. (ed.) Impacts of forest harvesting on long-term site productivity. Chapman & Hall, New York.

Karlen, D.L., and D.E. Stott. 1994. A framework for evaluating physical and chemical indicators of soil quality. p. 53–72. *In* J.W. Doran et al. (ed.) Defining soil quality for a sustainable environment. SSSA Spec. Publ. 35. SSSA and ASA, Madison, WI.

Karlen, D.L., M.J. Mausbach, J.W. Doren, R.G. Cline, R.F. Harris, and G.E. Schuman. 1997. Soil quality: A concept, definition, and framework for evaluation. Soil Sci. Soc. of Am. J. 61:4–10.

Karr, J.R. 1993. Measuring biological integrity: Lessons from streams. p. 83–104. *In* S. Woodley et al. (ed.) Ecological integrity and the management of ecosystems. St. Lucie Press, New York.

Kelting, D.L., J.A. Burger, S.C. Patterson, W.M. Aust, M. Miwa, and C.C. Trettin. 1999. Soil quality assessment in domesticated forests—A southern pine example. Forest Ecol. Manage. (In press).

Kimmins, J.P. 1996. The health and integrity of forest ecosystems: Are they threatened by forestry? Ecosyst. Health 2:5–18.

Kiniry, L.N., C.L. Scrivner, and M.E. Keener. 1983. A soil productivity index based upon predicted water depletion and root growth. Missouri Agric. Exp. Stn. Res. Bull. 1051. Columbia, MO.

Kozlowski, T.T., P.J. Kramer, and S.G. Pallardy. 1991. The physiological ecology of woody plants. Academic Press, New York.

Lal, R., and F.J. Pierce. 1991. The vanishing resource. p. 1–5. *In* R. Lal and F.J. Pierce (ed.) Soil management for sustainability. Soil and Water Conserv. Soc., Ankeny, IA.

Lal, R., and B.A. Stewart (ed.). 1995. Soil management: Experimental basis for sustainability and environmental quality. Advances in soil Science. Lewis Publ., Boca Raton, FL.

Larson, W.E., and F.J. Pierce. 1994. The dynamics of soil quality as a measure of sustainable management. p. 37–51. *In* J.W. Doran et al. (ed.) Defining soil quality for a sustainable environment. SSSA Spec. Publ. 35. ASA and SSSA, Madison, WI.

Monserud, R.A. 1984. Problems with site index: An opinionated review. p. 167–180. *In* J.G. Bockheim (ed.) Forest land classification: Experiences, problems, perspectives. Univ. of Wisconsin, Madison.

Montreal Process. 1995. Criteria and indicators for the conservation and sustainable management of temperate and boreal forests. Canadian Forest Serv. Catalogue Fo42-238/1995E. Canadian Forest Serv., Hull, Quebec.

Morris, L.A., and R.E. Miller. 1994. Evidence for long-term productivity change as provided by field trials. p. 41–80. *In* W.J. Dyck et al. (ed.) Impacts of forest harvesting on long-term site productivity. Chapman & Hall, New York.

Nambiar, E.K.S. 1996. Sustained productivity of forests is a continuing challenge to soil science. Soil Sci. Soc. Am. J. 60:1629–1642.

Noss, R.F. 1990. Indicators for monitoring biodiversity: A hierarchical approach. Conserv. Biol. 4:355–364.

Oldeman, L.R. 1994. The global extent of soil degradation. p. 99–118. *In* D.J. Greenland and I. Szabolcs (ed.) Soil resilience and sustainable land use. CAB Int., Oxon, England.

Olsen, S.R., and L.E. Sommers. 1982. Phosphorus. p. 403–430. *In* A.L. Page et al. (ed.) Methods of soil analysis. Part 2. 2nd ed. Agron. Mongr. 9. ASA and SSSA, Madison, WI.

O'Neill, R.V., W.F. Harris, B.S. Ausmus, and D.E. Reichle. 1975. A theoretical basis for ecosystem analysis with particular reference to element cycling. p. 28–39. *In* F.G. Howell et al. (ed.) Mineral cycling in southeastern ecosystems, Augusta, GA. 1–3 May 1974. ERDA Symp. Series (CONF-740513). Tech. Inform. Ctr., Office of Public Affairs, U.S. Energy Res. and Dev. Administration, Washington, DC.

Papendick, R.J., and J.F. Parr. 1992. Soil quality: The key to a sustainable agriculture. Am. J. Altern. Agric. 7:2–3.

Parr, J.F., R.I. Papendick, S.B. Hornick, and R.E. Meyer. 1992. Soil quality: Attributes and relationship to alternative and sustainable agriculture. Am. J. Altern. Agric. 7:5–11.

Powers, R.F., D.H. Alban, R.E. Miller, A.E. Tiarks, C.G. Wells, P.E. Avers, R.G. Cline, R.D. Fitzgerald, and N.S. Loftus, Jr. 1990. Sustaining site productivity in North American forests: Problems and prospects. p. 49–79. *In* S.P. Gessel et al. (ed.) Sustained productivity of forest soils. Proc. 7th North Am. For. Soils Conf., Vancouver, BC. 24–28 July 1988. Faculty of Forestry, Univ. of British Columbia, Vancouver.

Powers, R.F., A.E. Tiarks, J.R. Boyle. 1998. Assessing soil quality: Practicable standards for sustainable forest productivity inthe United States. p. 53–80. *In* E.A. Davidson et al. (ed.) The contribution of soil science to the development of and implementation of criteria and indicators of sustainable forest management. SSSA Spec. Publ. 53. SSSA, Madison, WI (this publication).

Pritchett, W.L., and W.H. Smith. 1974. Management of wet savanna forest soils for pine production. Fla. Agric. Expt. Stn. Tech. Bull. 762. Florida Agric. Exp. Stn., Gainesville.

Ramakrishna, K., and E.A. Davidson. 1998. Intergovernmental negotiations on criteria and indicators for the management, conservation, and sustainable development of forests: What role for forest soil scientists? p. 1–16. *In* E.A. Davidson et al. (ed.) The contribution of soil science to the development of and implementation of criteria and indicators of sustainable forest management. SSSA Spec. Publ. 53. SSSA, Madison, WI (this publication).

Roberts, J.A., W.L. Daniels, J.C. Bell, and J.A. Burger. 1988. Early stages of mine soil genesis in a Southwest Virginia spoil lithosequence. Soil Sci. Soc. Am. J. 52:716–723.

Salwasser, H., D.W. MacCleery, and T.A. Snellgrove. 1993. An ecosystem perspective on sustainable forestry and new directions for the U.S. National Forest System. p. 44–89. *In* G.H. Aplet et al. (ed.) Defining sustainable forestry. Island Press, Washington, DC.

Singh, K.K., T.S. Colvin, D.C. Erbach, and A.Q. Mughal. 1992. Tilth index: An approach to quantifying soil tilth. Trans. ASAE 35:1777–1785.

Smith, J.L., J.J. Halvorson, and R.I. Papendick. 1993. Using multiple variable indicator kriging for evaluating soil quality. Soil Sci. Soc. Am. J. 57:743–749.

Stone, E.L. 1975. Soil and man's use of forest land. p. 1–9. *In* B. Bernier and C. Winget (ed.) Forest soils and forest land management. Laval Univ. Press, Quebec.

Terry, T.A., and J.H. Hughes. 1975. The effects of intensive management on planted loblolly pine growth on poorly drained soils of the Atlantic Coastal Plain. p. 351–377. *In* B. Bernier and C. Winget (ed.) Forest soils and forest land management. Laval Univ. Press, Quebec.

Torbert, J.L., J.A. Burger, and W.L. Daniels. 1989. Pine growth variation associated with overburden rock type on a reclaimed surface mine in Virginia. J. Environ. Qual. 19:88–92.

Wagenet, R.J., and J.L. Hutson. 1997. Soil quality and its dependence on dynamic physical processes. J. Environ. Qual. 26:41–48.

World Commission on Environment and Development. 1987. Our common future. Oxford Univ. Press, Oxford, England.

3 Assessing Soil Quality: Practicable Standards for Sustainable Forest Productivity in the United States

Robert F. Powers

Pacific Southwest Research Station, USDA Forest Service
Redding, California

Allan E. Tiarks

Southern Research Station, USDA Forest Service
Pineville, Louisiana

James R. Boyle

Department of Forest Resources
Oregon State University
Corvallis, Oregon

Productive soils form the foundation for productive forests. But unfortunately, the significance of soil seems lost to modern society. Most of us are too far removed from the natural factors of production to appreciate the multiple roles of soil. Nor is its worth recognized well by many forest managers who too often see soil only in its capacity for logging roads and trafficability. Thus, the Santiago Declaration is a milestone marking the significance of soil in sustainable forestry.

Soil-based indicators of sustainable forestry listed in the Santiago Declaration are described in a generic sense, and countries are left with the task of developing objective standards of soil quality. Here, we examine the historical and conceptual basis for soil quality standards as proposed by the Santiago Declaration relative to U.S. forestry. Further, we synthesize findings from recent research and suggest an approach to develop operationally practicable standards of soil quality for managed forests of the nation.

PERCEPTIONS

Historical Views in the USA

Measured by our lifespans, soils are nonrenewable resources—something to be protected, restored where necessary, perhaps enhanced. But this was not always the view in the USA. In 1909, the Chief of the U.S. Bureau of Soils wrote,

The Contribution of Soil Science to the Development of and Implementation of Criteria and Indicators of Sustainable Forest Management. SSSA Special Publication no. 53.

"The soil is the one indestructible, immutable asset that the nation possesses. It is the one resource that cannot be exhausted, that cannot be used up" (Whitney, 1909). Fortunately, this sanguine view has not prevailed.

The westward expansion triggered by the Homestead Act of 1862 began a period of soil exploitation in the USA, culminating in boom-and-bust agricultural production during and following World War I. Falling markets, the Great Depression, and prolonged periods of inclement weather led to extensive erosion and declining soil fertility throughout much of our farmland. Largely through the dramatic writings of Hugh Hammond Bennett, Director of the Soil Erosion Service (who once stalled a congressional hearing until a dust cloud arrived, blotting out the sun), Congress was moved in 1935 to pass the Soil Conservation Act and establish a national Soil Conservation Service. The Act was landmark U.S. legislation because it highlighted soil as a heritage to be protected, repaired, or improved. The Act established that it was in the public interest to do so.

Principles of soil conservation known for centuries were given new life in Bennett's extensive 1939 treatise. But applying such principles requires a long-term perspective. Paybacks from soil conservation or rehabilitation practices often accrue so slowly that their long-term worth is not evident. Lacking other criteria or indicators, landowners turn to short-term profits to weigh new practices. This tends to favor conventional practices, even if they may be detrimental in the long run. Consequently, many conservation or rehabilitation practices are not adopted or are abandoned prematurely. Clearly, the first step toward improved soil management lies in convincing managers that change is needed.

In the presence of apathy or ignorance, it falls upon government to protect our natural resources and their potential values for the good of future generations. Australia, the Netherlands, and the USA stand alone among the world's nations in their legal mandates for soil protection and land stewardship. The 1992 National Forest Policy Statement by the Australian Government holds that forest management strategies and operations must ensure protection of ecological processes, soil quality, and sustained yield (Nambiar, 1996). The Dutch Soil Protection Act of 1987 requires that soil of the Netherlands cannot be treated in a way that degrades its capacity for such multiple functions as grazing, groundwater recharge, or crop production (Moen, 1988). In the USA, the Multiple Use and Sustained Yield Act of 1960, the National Environmental Policy Act of 1969, the Forest and Rangeland Renewable Resources Planning Act of 1974, and the National Forest Management Act of 1976 (NFMA) all bind the USDA Forest Service to manage public lands without damaging their permanent productivity (USDA Forest Service, 1983). The Forest Service interprets permanent productivity of the land as the capacity of a site to sustain forest growth—a capacity depending largely on the quality of the soil (Powers & Avers, 1995).

The U.S. laws, of course, apply to federal land, but not to state or private forests. For nonfederal lands, fewer than one-half of the states have adopted or are considering adoption of forest practice regulations designed to protect forest values. Regulations applying to soil concern erosion and its implications for water quality, but seldom address other soil properties or processes. In states without forest practice rules, apprehension of regulation as much as altruism has spurred many companies to self-impose best management practices aimed at securing reforestation, maintaining water quality, and enhancing wildlife values. But often these practices seem more of a goal than a rule.

Environmental sensitivity is now a marketable factor, and a variety of national and international programs have surfaced that purport to certify sustainable forestry according to standards they have adopted (Journal of Forestry, 1995). Many in forest industry are skeptical of third-party green certification where criteria may be based on arbitrary and unscientific standards (Berg & Olszewski, 1995). In response, the American Forest and Paper Association (AF&PA) has adopted a set of sustainable forestry principles and guidelines to ensure that industrial forestry lands are managed to meet or exceed the standards called for in federal regulations (AF&PA, 1994). Member companies are free to establish their own policies and plans to achieve AF&PA principles. But guidelines are quite broad and apply only tangentially to soil. We believe that forest managers in both public and private sectors wish to be good land stewards. But managers are not apt to alter their practices without a compelling reason.

Soil Quality Concept

In 1992, a special symposium of the American Society of Agronomy met in Minneapolis, MN, to address the theme: "Defining Soil Quality for a Sustainable Environment." The publication that followed (Doran et al., 1994) defined soil quality as "the capacity of a soil to function within ecosystem boundaries to sustain biological productivity, maintain environmental quality, and promote plant and animal health" (Doran & Parkin, 1994). This definition frames the soil quality concept. It's application pivots on two questions (Karlen et al., 1997): (i) How does the soil function? and (ii) What indicators are best for evaluation?

Unfortunately, soil quality cannot be measured directly. The trick is to define the functional elements of soil that sustain biological productivity and to identify soil quality indicators of these functions. Useful indicators must be sensitive to variations in management and climate, integrate soil physical, chemical and biological properties and processes, and be practicable on a variety of scales by a variety of users. Warkentin (1995) discussed the need for flexible and adaptable views of soil quality to suit various purposes. Larson and Pierce (1994) listed 10 key soil attributes as a minimum data set for monitoring quality. These included nutrient availability, available water capacity, and measures of soil structure and strength. Doran and Parkin (1994) proposed a primary set of 16 soil quality indicators related to these attributes. These ranged in complexity from soil depth to ratios of soil respiration to microbial biomass. This primary set could be supplemented to provide more detail. The context was North American agriculture. This approach has merit for agricultural systems with less spatial variability and more management intensity than found commonly in forests (but see Burger & Kelting, 1998, this publication, for application to forestry). For operational forestry, we suggest an alternative approach.

EXTENDING THE CONCEPT TO FORESTRY

We agree that forest soil quality is the relative capacity of a soil to sustain the processes and produce the properties critical to ecosystem function. But processes and properties of forest soils are more varied and less understood than those characterizing modern agriculture. North American forestry runs a broader gamut of technological intensities than domestic agriculture (Stone, 1975). Except for the industrial Southeast and portions of the coastal Pacific Northwest, most forests are not fully regulated, and some management could be considered exploitative. Forest conditions vary from virgin old-growth forests of irregular structure and minimal management to multigeneration clonal plantations growing on abandoned farm lands. The simplest and most direct reflection of the quality of this broad range of sites, relative to sustainable forestry, is their inherent capacity to sustain the growth of vegetation (measured as total net primary productivity) when sites are at leaf area carrying capacity. We define this as *inherent site productivity* (Powers, 1999).

Inherent site productivity is a function of climate, biotic potential, and soil. As is true for agriculture, site productivity of managed forests should not be seen as a ceiling. Instead, it is a potential that can be raised or lowered by management. But assessing soil productivity—the portion of inherent site productivity attributable to soil—is not simple. Our conventional forestry measure of productivity is tree growth, but tree growth is confounded easily by weed competition—particularly in the early stages of stand development. Should the standard be dry matter production at a later stand age as measured in consecutive rotations? Ratios of yields in past, present, and future rotations, an attractive mensurational approach, can be misleading and even irrelevant because of changes in climate, genotype, silviculture, and human aspirations (Vanclay, 1996). Growth gains due to improvements in genotype and stocking can easily mask early declines in soil productivity, particularly in short rotations (Powers et al., 1996). Thus, assessing soil productivity in forestry is more difficult than in modern agriculture where slopes are gentle, stone content is low, surface soils have been homogenized by tilling, and crop yields from fully stocked fields generally can be measured annually.

There are other formidable problems in applying the soil quality concept in forestry. North America's forest soils are extremely diverse. They encompass essentially all of the soil taxa defined by the U.S. and Canadian classification systems, and much forest land awaits classification. Climates vary from the mild tropical conditions of Hawaii to the harsh, continental conditions of the Alaskan taiga. Complexities caused by all of the above make the basic assessment of soil productivity particularly difficult. Soil testing methods popular in agronomy and commonly used in assessing soil quality of agricultural soils have been marginally helpful in forestry. We see little justification for adopting analytic methods lacking sound correlations with the growth behavior of forest vegetation. Also, spatial variability creates an immense problem for estimating site means with reliability and precision. Therefore, we do not believe that the detailed, intensive approach of Doran and Parkin (1994) is justified for monitoring forest soil quality extensively in the USA (but see Burger & Kelting, 1998, this publication, for an application to intensive forestry).

Despite the problem of assessing forest soil quality, we do agree with the concept and with the principle of soil quality monitoring. Sustaining or enhancing soil quality is the foundation for sustained forest productivity and periodic assessment seems entirely logical. But soil quality monitoring programs should be based on substance. Is there convincing evidence that soil quality standards are warranted or feasible in forestry?

Is Soil Quality Declining from Forest Management?

Morris and Miller (1994) and Powers et al. (1990) have reviewed the apparent evidence that soil productivity has declined under forest management. Lessened growth often traces to causes other than soil disturbance. For example, findings from repeated forest inventory in Georgia show a progressive decline in diameter growth of pine between 1956 and 1982 (Sheffield et al., 1985); however, declines were restricted to nonindustrial private forest land where shrub and hardwood competition had increased from the absence of repeated underburning. On more intensively managed industrial plantations, growth rates of pine were stable or had increased (Sheffield & Cost, 1986).

Conversely, effects of disturbances likely to degrade the soil can be masked by confounding factors such as differential weed competition. Comparing site preparation techniques, Stransky et al. (1985) found that both survival and growth of loblolly pine (*Pinus taeda* L.) through 8 yr were greatest where surface materials had been bladed by tractor into windrows. Weed competition often is lessened by windrowing if the seed banks and sprouting roots present in topsoil are windrowed as well. The fundamental effect is a redistribution of soil fertility. Early benefits of windrowing in the Stransky et al. (1985) study masked a more significant loss in potential soil productivity from sizable losses in soil organic matter. Similarly, Allen et al. (1991) found that loblolly pine had higher survival, less weed competition, and grew twice as fast where slash had been windrowed than where slash had been chopped and burned. But when weed and tree biomass were combined, total productivity was greatest in the chop and burn treatment. By year 8, pine height growth was best where slash had been chopped and burned. In both studies, superior early growth was due to enhanced tree survival and better weed control in the most severe treatment.

Our tendency to focus only on tree growth misses the fact that total productivity of all vegetation often is greatest on the least severe treatment. Such narrow focus can lead managers to repetitive practices that degrade the soil until it is obvious to everyone. Problems may not appear for several years because nutrient releases from root decomposition, plus nutrients already available in the mineral soil, often exceed the needs of very young vegetation (Smethurst & Nambiar 1990). Burger and Kluender (1982) concluded that windrowing practices favoring early growth in southern pine plantations would lead to growth declines by year 10.

Forests accumulate immense amounts of organic N and P in their standing biomass and forest floor, and substantial amounts can be removed during logging and site preparation. Yet, direct evidence is rare that nutrient removals in biomass harvesting trigger declines in soil productivity. Perhaps the best-known case in North America is the small plot experiment of Compton and Cole (1991) with Douglas-fir (*Pseudotsuga menziesii*). Three residue treatments were applied to small plots following the clearcutting of natural stands on two sites of contrasting quality. Removal treatments were (i) stem only, (ii) whole tree, and (iii) whole tree plus understory and forest floor removal. On the poorer site, these treatments removed 16, 31, and 51% of the ecosystem N to a soil depth of 50 cm. Plots were then planted with Douglas-fir. Tree heights after 10 yr were inversely proportional to the amount of organic matter removed on the poorer site (results were erratic on the better site). Fertilizing one replicate of each treatment with N at 5 yr increased growth rates in proportion to organic matter removed, suggesting that removing organic matter induced N deficiency. Small plot size and failure to account for competing vegetation limit the utility of this study, but results do show a correspondence between organic matter removal and lessened tree growth. The authors concluded that biomass removal had a direct effect on soil productivity by altering the N cycle.

Most U.S. findings are short-term or inconclusive, and have centered on nutritional questions. But other cases offer strong arguments that soil quality is degraded by management practices affecting soil physical properties. Tiarks and Haywood (1996) compared first- and second-rotation responses of slash pine (*Pinus elliotii*) to three residue treatments on silt loam soils in Louisiana. Treatments were established in 1960 after a natural stand of pines and hardwoods was harvested and the logging slash burned. Treatments were applied when soils were moist and consisted of (i) no further treatment (burn-only), (ii) disking, and (iii) disking plus bedding. Trees were planted and heights were measured yearly until first-rotation harvesting in 1983. Slash was burned again on all plots, but mechanical site preparation was not repeated. The plots were replanted in 1984 with slash pine believed to be superior genetically to that in the first rotation and weeds were controlled in all treatments.

Analysis of periodic height growth through the first rotation showed that trees were substantially taller in the disking plus bedding treatment. But in the second rotation, the burn-only treatment had consistently greater rates of height growth than either of the mechanical site preparation treatments, and the disking treatment had the poorest growth rate of all. Soil strength profiles measured in the second rotation, 33 yr after the original site preparation, indicated large, continuous increases in strength below 20 cm in disked plots, and a similar but discontinuous band in bedded plots (Fig. 3–1). Mechanical site preparation produced high yields in the first rotation, but lower yields in the second. Results from an auxiliary study suggest that incipient P deficiency may have been exacerbated as well, particularly where the compacted traffic pan was continuous. The fact that weed competition was negligible among treatments indicates that tree growth was a good reflection of soil productivity. From this, it follows that soil productivity was degraded in the long term by cumulative physical impacts of mechanical site preparation.

Such anecdotal findings from the USA and throughout the world suggest that management practices can reduce soil productivity through declines in site organic matter and soil physical properties (Powers et al., 1990). Linkages between soil physical properties and organic matter are shown by the conceptual model in Fig. 3–2. Changes in a physical property such as soil porosity alter water infiltration and runoff at the soil surface and the exchange of water and gasses at depth. Changes in soil strength and soil aeration affect root respiration and the penetration of soil voids by roots. Thus, the volume of soil available for rooting

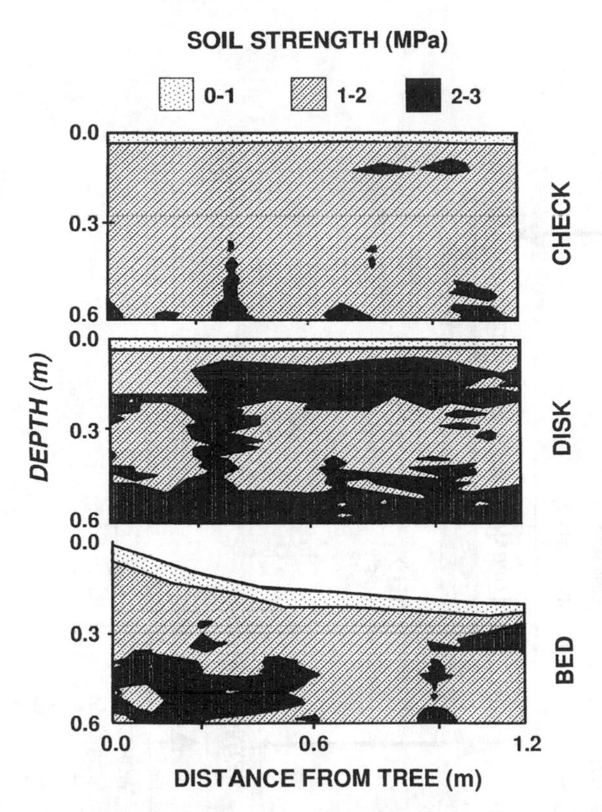

Fig. 3–1. Penetrometer measurements of soil strength on burned only (Check); burned and disked (Disk); and burned, disked and bedded (Bed) treatments measured 33 yr after site preparation. Measurements were taken at 10-cm intervals away from rows of planted trees (Tiarks & Haywood, 1996).

will be altered. These same changes affect water and nutrient flow to roots as well as the activity of microbes involved in organic matter decomposition and nutrient mobilization. Site organic matter also provides a nutrient-rich substrate important to the diet of soil fauna. Changes in their activity lead to parallel changes in organic matter comminution and the enhancement of substrates for bacteria and fungi. This alters the dynamics of soil structure, as well as the release rate of nutrients to plants.

It follows from this that changes in soil physical properties and organic matter content alter soil quality by changing the ability of roots to support leaf mass and the primary productivity of a site. The likelihood that such soil properties are degraded by management is greatest during timber harvest and site preparation. The significance is the ripple effect of soil disturbance on more fundamental properties and processes affecting root activity.

Are Soil Disturbance Effects Universal?

How far can anecdotal evidence extend? Are universal guidelines possible? If the loss of appreciable surface organic matter leads invariably to site decline,

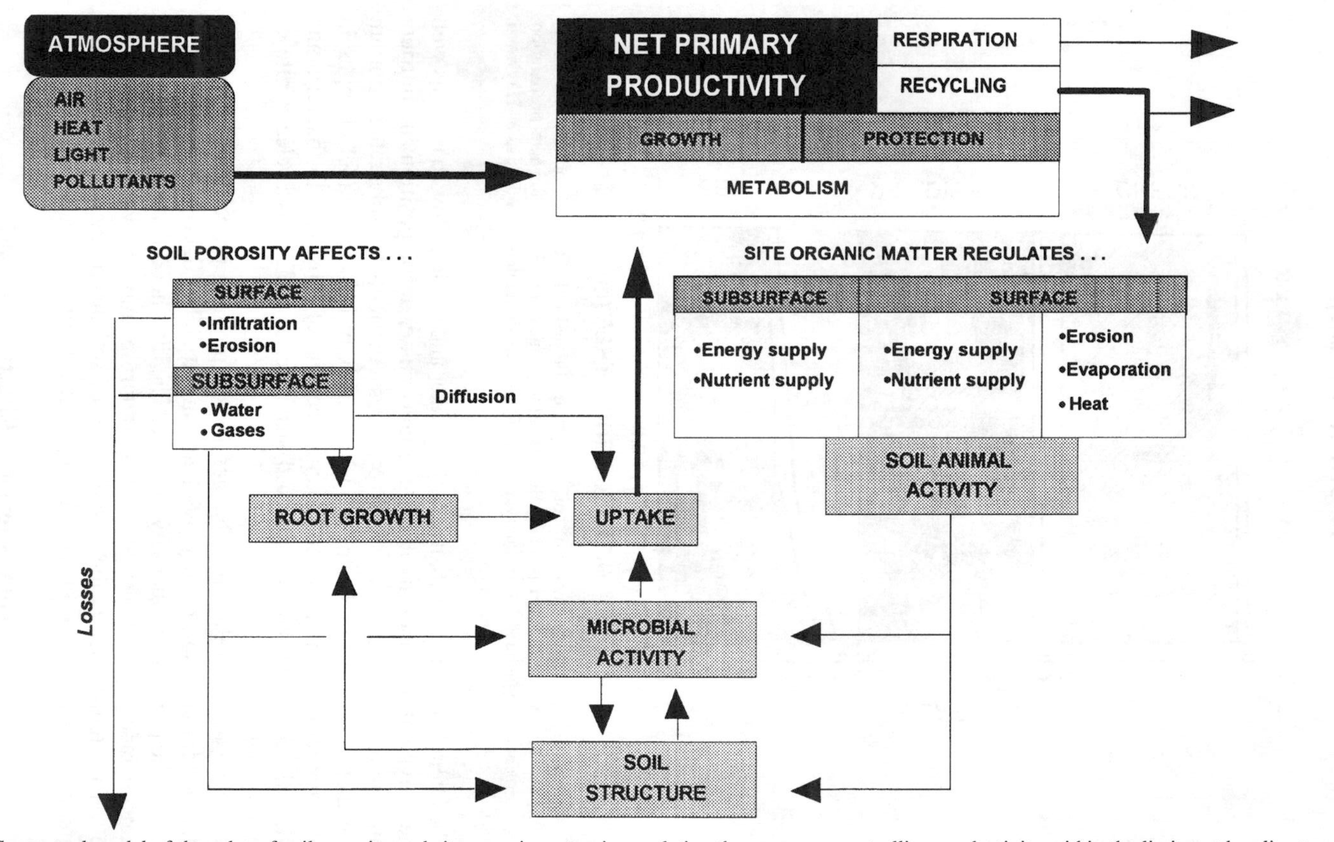

Fig. 3–2. Conceptual model of the roles of soil porosity and site organic matter in regulating the processes controlling productivity within the limits set by climate and genotype (Powers et al., 1990).

Table 3–1. Management effects on soil organic matter (OM) and key soil processes affecting long-term net primary productivity (L-T NPP) in temperate and boreal forests. Effects are positive (+), negative (-), none (0), or unknown (?).

Event	OM input to soil	Soil temper-ature	Soil moisture	Soil biotic activity	Aggreg. stability	Nutr. availability	Root activity	L-T NPP
Harvesting								
Logs removed	–	+	+	0	0	0	0	0
Slash produced	+	+	+	0	0	0	0	0
Site preparation								
Slash retained	+	–	+	+/–	+	+/–	+/–	+/–
Slash tilled	+	?	+	+	?	+	+	+
Slash removed	–	+	–	+/–	–	+/–	+/–	0/–
Vegetation control	–	+	+	–	–/?	+/–	+	+/–
Subsoiling	+	?	+	+	+/?	+	+	+/?
Fertilization	+	–	–	+/–	+	+	+/–	+
Irrigation	+	–	+	+	+	+	+	+

establishing soil quality indices and standards would be simplified. We doubt this is the case. Table 3–1 illustrates general cause and effect linkages between common forest management practices and potential site productivity as measured by net primary productivity when the forest is at leaf area carrying capacity. Certain operations have a consistent effect on soil properties. For example, conventional tree harvesting adds organic matter as a pulse to the soil surface. Loss of canopy leads to warmer summer soil temperatures, to greater diurnal heat flux, and to greater soil moisture from reduced transpiration. These are universal facts.

But the effects of this on biological processes depend on other factors. If logging slash is retained, its insulating properties reduce surface soil temperature and evaporative losses of soil moisture. A soil that is cooler and moister may promote microbial activity in warm, dry climates, but has an opposite effect in the boreal forest. In British Columbia, summer soil temperatures are raised as much as 4°C when forest floors and competing vegetation are removed (Fleming et al., 1994; Messier, 1993). Soil warming in higher-latitude forests stimulates decomposition, N and P mobility, and nutrient uptake (Van Cleve et al., 1990). On warm, dry sites, mineralization is reduced (Powers, 1990).

In the short run, root activity and primary productivity should increase following removal of surface organic matter on cold, wet sites, but will decrease on hot, dry sites. In the long term, benefits of pulse removals of logging slash will dissipate, and fertility could decline to a biologically significant degree on infertile sites. Similar analyses could be made for such common management practices as vegetation control, tillage, and fertilization. For example, eliminating understory vegetation, a common practice to improve moisture and nutrient availability and stimulate tree growth on many sites (Messier, 1993; Powers & Ferrell, 1996), can significantly reduce rates of soil C and N accumulation if done repeatedly (Busse et al., 1996). And fertilizing N-deficient forests with biosolids may either promote or depress tree growth, depending on the C/N ratio of the substrate and the cation status of the soil (Harrison et al., 1996). Shallow tilling of fine textured soils under moist conditions may benefit growth in the current rotation, but lead to a compacted plow pan in subsequent rotations (Tiarks & Haywood, 1996).

Rarely do rules-of-thumb apply to the myriad of conditions characterizing forests of the USA. Universally, soil condition and climate must be considered in judging the consequences of forest management practices.

Climate and soil condition also influence the significance of physical soil changes on productivity. The first attempt at generalizing the impact of soil compaction on forest growth was the empirical model of Froehlich and McNabb (1984) that predicted a 6% reduction in seedling height growth for every 10% increase in soil bulk density. Although a good first step, this linear model is unrealistic because it does not account for changes in soil aeration capacity. Reicosky et al. (1981) showed that increasing soil bulk density of a loam from 1.0 to 1.2 Mg m^{-3} (an increase of 20%) reduced macropore volume by 39%. But this reduction in the volume of very large pores also increased low-tension water volume by 17%, which could promote growth during drought. Raising bulk density further from 1.2 to 1.3 Mg m^{-3} (an increase of only 8%) reduced aeration capacity below the standard of 0.1 m^3 m^{-3} considered vital for root respiration (Vomocil & Flocker, 196; Grable & Siemer, 1968). Raising bulk density from 1.3 to to 1.6 Mg m^{-3} (an increase of 23%) reduced macroporosity another 60% and lowered the volume of low-tension water by 40%. Clearly, potential biological impacts of soil compaction are not linear, but depend in large measure on pore-size distribution and aeration porosity.

Beside its effect on aeration porosity, compaction also must be interpreted by its effect on soil strength. This is shown by the greenhouse work of Simmons and Pope (1988) using hardwood seedlings and a silt loam soil compacted to different densities (Fig. 3–3). Under moist conditions (ψ = −10 kPa), increasing bulk density by 24% from 1.25 Mg m^{-3} to 1.55 had little effect on soil strength but air-filled pore volumes were reduced below 0.1 m^3 m^{-3}. This created anaerobic conditions, and root growth nearly ceased. Under drier conditions (ψ = −300 kPa), aeration porosity was adequate at all densities, but soil strength restricted root growth at the highest bulk density. The net effect of compaction was restrict-

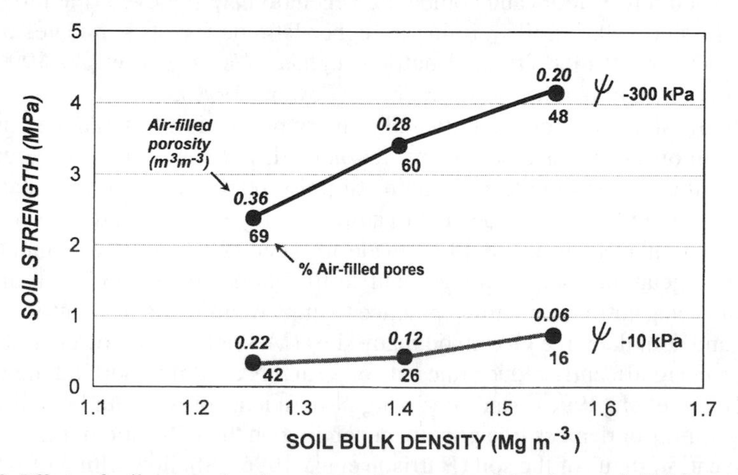

Fig. 3–3. Effects of soil compaction depend on soil water potential and strength. Holding soil water potential at −10 kPa, compaction to 1.55 Mg m^{-3} creates anaerobic conditions. At lower water potential, soil strength becomes limiting to root growth (Simmons & Pope, 1988).

ed root penetration into new soil volumes because of poor aeration under moist conditions and high particle resistance under drier conditions.

In a field study of logging in coastal Washington, Miller et al. (1996) found that increased soil bulk density on loam to clay loam soils had no discernible impact on Douglas-fir growth after nearly two decades. They attributed this to initially low bulk densities, the high organic matter content of the soil, and the mild coastal climate that minimized drought stress. In southeastern British Columbia, soil compaction associated with skidroads reduced the growth of planted conifers on a silt loam, but changes were less severe—and often improved—on a sandy loam (Senyk & Craigdallie, 1997). Powers (1997) studied reasons for improved conifer seedling growth on a sandy-textured soil following moderate compaction. A decrease in macroporosity increased available water-holding capacity, translating to higher seedling water potential and a volume growth increase of 60%. In contrast, compacting a clay loam decreased available water-holding capacity, seedling water potential, and growth. We conclude that simple models such as that by Froehlich and McNabb (1984) are ineffective because their single variable, bulk density, has limited biological relevance. More useful models would account for aeration, moisture availability, and strength.

Physical traits also affect tree response to soil disturbance. In a factorial experiment of soil compaction and organic matter removal following clearcutting on a loamy sand in Minnesota, Alban et al. (1994) studied second-year production of aspen (*Populus tremuloides* and *P. grandidentata*) sprout regeneration. As expected, soil bulk density and strength were increased and infiltration was decreased significantly with compaction. The aboveground biomass of aspen regeneration was reduced 44% by compaction and 34% by removing the forest floor. But the combined treatments stimulated suckering so that biomass productivity declined by only 14% through the first 2 yr.

Forest responses to soil disturbance clearly are not universal, but depend on interactions among climate, soil condition, and certain physiological traits. Therefore, searching for specific soil quality indicators (e.g., base saturation, bulk density, and organic matter content) with universal meaning seems futile. Our focus should be at a broader and more integrative level reflecting the dominant processes important to long-lived vegetation. The nature of forest management requires that whatever the approach, it must be affordable for extensive, operational forestry.

United States Forest Service Approach

The USDA Forest Service emphasizes soil quality standards in forest management planning. Each of the nine Forest Service administrative regions has developed monitoring standards meant to detect changes that would lower soil productivity over a rotation. Changes beyond established thresholds indicate the need for mitigation. Recognizing that productivity standards cannot rely solely on tree growth, the Watershed and Air Management staff of the Forest Service adopted a program for monitoring soil productivity that is based on the following logic: (i) management practices create soil disturbances; (ii) soil disturbances affect soil and site processes; and (iii) soil and site processes control site productivity.

Monitoring soil and site processes is not operationally feasible. For example, the supply and release of organically bound nutrients is central to sustaining soil quality; however, net mineralization rates depend on soil moisture, temperature, and the nature of the organic substrate (Adams & Attiwill, 1986; Powers, 1990), making process monitoring too costly and complex to be practical. Instead, Forest Service monitoring strategy focuses on measurable soil disturbance variables such as compaction, ground cover, soil displacement, and organic matter abundance that influence important site processes. Soil management staff of each Forest Service region then develop independent standards for each variable that mark thresholds for detrimental soil disturbance. A threshold for detrimental disturbance is defined as a change in any monitoring variable sufficient to trigger a 15% decline in soil productivity from that of the predisturbed condition, a value judged to be the smallest change detectable statistically at operational levels of monitoring. This does not imply that absolute productivity has declined 15%, but merely that a detrimental disturbance threshold has been passed.

In practice, operational areas are surveyed following an activity such as logging and characterized by the amount of detrimental soil disturbance. If the sum of detrimental disturbance of all types on an operational area exceeds a defined areal threshold (generally, 15% of the area capable of growing trees), mitigative measures are taken. Ideally, variables and threshold values are based on scientific findings. Lacking this, they are based on best professional judgement. Either way, they must be operationally practicable and open to continuing revision. Standards as of this writing are summarized in Table 3–2.

Table 3–2. Soil quality standard threshold for various disturbance variables for the nine administrative Forest Service Regions (1, Northern; 2, Rocky Mountain; 3, Southwestern; 4, Intermountain; 5, Pacific Southwest; 6, Pacific Northwest; 8, Southern; 9, Eastern; 10, Alaska.

Disturbance variable	FS region	Threshold value
Operational area	1	The total of all detrimental conditions should not exceed 15% of the activity area, exclusive of system roads.
	2	The total of all detrimental conditions should not exceed 15% of the activity area, exclusive of system roads.
	4	The total of all detrimental conditions should not exceed 15% of the activity area, exclusive of roads, dedicated trails, mining excavations,and dumps.
	5	Standards apply to all land dedicated to growing vegetation.
	6	The total of all detrimental conditions should not exceed 15% of the activity area, exclusive of roads and landings.
	8	The total of all detrimental conditions should not exceed 15% of the activity area, exclusive of roads and landings.
	9	The total of all detrimental conditions should not exceed 20% of the activity area.
	10	The total of all detrimental conditions should not exceed 15% of the activity area.

(continued on next page)

Table 3–2. Continued.

Disturbance variable	FS region	Threshold value
Altered wetness	10	Areas become perrenially flooded and drained, and the natural function and value of the land is lost.
Erosion (surface)	1	See soil cover.
	2	See soil cover.
	3	Current soil loss exceeds tolerance soil loss. Loss of 75 to 100% of the A horizon. Rills and gullies connected and expanding.
	4	See soil cover.
	5	Soil loss should not exceed the rate of soil formation, or about 2 Mg ha^{-1} yr^{-1} as determined by Regional standards.
	6	See soil cover.
	8	Soil loss exceeds the allowable loss tolerance values set by the Region 8 Guide.
	9	Sheet and rill erosion exceeds the average annual soil loss tolerance over a rotation, or exceeds twice the threshold.
	10	Removal of 50% of topsoil or humus-enriched surface soil from 9.3 m^2 or more.
Soil cover	1	Enough cover to prevent erosion from exceeding natural rates of formation.
	2	Depending on erosion hazard class, effective ground cover is <30 to 50% the first year, and 50 to 70% the second year.
	4	Too little to prevent erosion from exceeding natural rates of soil formation determined through the Universal Soil Loss Equation.
	5	Forest floor covers <50% of area.
	6	Less than 20% cover on sites with low Erosion Hazard Ratings, 30% for moderate, 45% for high, and 60% for very high in first year after disturbance. Standards rise to <30, 40, 60, and 75% in the second year.
	8	Cover guided by local standards.
	10	Less than 85% litter retention on slopes <35%. Less than 95% cover on slopes >35%.
Organic matter	1	Less than 30% litter retention. Large woody debris based on Habitat Type.
	2	Less than 90% retention of fine logging slash following clearcutting or seedtree cutting on soils rated severe. Less than 50% retention following shelterwood or group selection cuttings.
	3	Litter absent or not evenly distributed. Large woody debris less than 11 to 16 Mg ha^{-1}, depending on habitat type.
	4	Large woody debris and litter insufficient to sustain site productivity determined through research. Standards depend on local conditions.
	5	Litter and duff cover <50% of area. Fewer than 12 decomposing logs ha^{-1} with diameters of at least 50 cm and lengths of 3 m.
	6	See soil displacement.
	8	Soil organic matter <85% of that in upper 30 cm of undisturbed soil, or losses great enough to degrade the nutrient cycle.
Infiltration	1	Reduction of 50% in natural rate.
	3	Reduction of >50%.
	5	Reduced to ratings of 6 or 8 as defined by Regional Erosion Hazard Rating System. Extent depends on cumulative watershed effects analysis.
Compaction	1	Bulk density increase of 15% (20% on volcanic soils).
	2	Bulk density increased >15% over natural conditions, or exceeding 1.25 to 1.60 g cm^{-3}, depending on soil texture.

(continued on next page)

Table 3–2. Continued.

Disturbance variable	FS region	Threshold value
	3	Bulk density increased by >15%. Soil strength increased by >50%. Massive or platy structure or lack of tubular pores in surface soil.
	4	Reduction of >10% in total soil porosity or a doubling of soil strength in any 5-cm increment of surface soil.
	5	Reduction of >10% in total soil porosity over an area large enough to reduce productivity potential. Allowable change should be determined for each soil.
	6	15% bulk density increase, 50% macropore reduction from the undisturbed, and/or a 15% reduction in macropore space measured by air permeameter for soils other than Andisols; 20% increase in bulk density for Andisols or Andic materials.
	8	15% bulk density increase and >20% decrease in macroporosity over undisturbed conditions.
	9	15% bulk density increase over undisturbed conditions.
	10	15% bulk density increase over undisturbed conditions.
Rutting and puddling	1	Soil puddling present.
	4	Ruts or hoof prints in mineral soil or Oa horizon of an organic soil.
	6	Ruts to at least 15 cm depth.
	8	Ruts exceed 15 cm deep for a continuous distance of >15 m, ruts exceed 30 cm deep for >3 m, and ruts exceed 46 cm for any distance.
	9	Ruts exceed 46 cm deep anywhere on site, or rut depths exceed 30 cm and extend for >3 m.
	10	Ruts or hoof prints in mineral soil or Oa horizon of an organic soil.
Detrimental burning	1	Loss of O horizon and signs of mineral soil oxidation.
	2	Most woody debris and the entire forest floor consumed to bare mineral soil. Soil may be reddened. Fine roots and organic matter charred in upper 1 cm of mineral soil.
	3	See organic matter.
	4	Loss of >5 cm or one-half of naturally occurring litter layer, whichever is less.
	6	Top layer of mineral soil changed in color to red and next 1 cm blackened from charring.
	9	Consumption of the forest floor to mineral soil on an area of >4.6 m^2. Bare soil exposed.
	10	Most woody debris and entire forest floor consumed to bare mineral soil and fine roots charred in upper 1 cm of mineral soil.
Displacement	1	Loss of either 2.5 cm or more of the surface soil, or one- half of the humus-enriched A horizon, whichever is less.
	2	Soil loss from a continuous area of >9 m^2.
	4	Removal of >50% of humus-enriched surface soil or 5 cm of soil from an area at least 1 m^2.
	5	Organic matter in the upper 30 cm of soil is <85% of the soil organic matter found under natural conditions. Affects an area sufficiently large that productivity potential is reduced.
	6	Removal of >50% of topsoil or humus-enriched A1 and/or AC horizons from an area of 9.3 m^2 or more and at least 1.5 m wide.
	8	Removal of >50% of the humus-enriched A horizon over a continuous area >5.6 m^2 and >1 m wide.
	9	Removal to a depth of one-half the thickness of the A horizon over an area of >5.6 m^2 and >1m wide.
	10	Removal of forest floor and 50% of topsoil from an area of 9.3 m^2 and least 1.5 m wide.

Soil quality standards of the USDA Forest Service are meant as early warning thresholds of impaired soil conditions. Figure 3–4 illustrates the concept. For a particular soil and site, a change in a key soil monitoring variable will lead ultimately to a change in potential productivity. For the hypothetical model in Fig. 3–4, the change is negative. Obviously, not all soil changes caused by management are detrimental. Soil productivity may be favored by fertilization on nutrient-deficient sites, or by improving soil moisture regimes on overly wet or dry soils. Because of the stewardship requirements of NFMA (USDA Forest Service, 1983), the Forest Service focuses only on those variables thought to be associated with detrimental change. The critical feature of the concept is that the soil monitoring variable must be linked very closely with potential productivity when the site is fully stocked with vegetation. Unfortunately, calibrations of such linkages (Fig. 3–4) do not exist.

As a remedy, the Forest Service began a national long-term soil productivity (LTSP) cooperative effort in 1989 to validate the operational regional standards, to identify new, more effective monitoring variables, and to develop calibration curves for major soil and forest types of the USA (Powers et al., 1990; Powers & Avers, 1995). To date, more than 60 installations with a common factorial experimental design and measurement protocol exist in the USA and Canada, and many studies with common elements and goals are affiliated with LTSP (Powers et al., 1996). All LTSP installations are <10 yr old as of this writing. Until they mature, interim standards shown in Table 3–2 will be used as thresholds that, according to best professional judgement, indicate when detrimental soil disturbance has occurred.

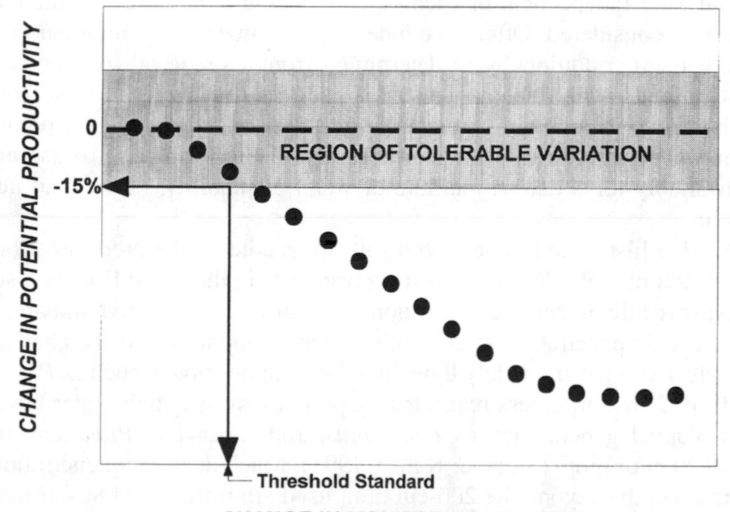

Fig. 3–4. Hypothetical relationship between potential productivity and a key soil variable. In principle, as properties of a soil variable degrade, productivity declines from its potential. In practice, variability exists about the estimate of potential productivity for an undisturbed site. Threshold standards are meant to indicate the soil quality change equivalent to a 15% decline in potential productivity.

The selection of disturbance variables common to most or all Forest Service regions shows consensus that surface erosion, compaction, and displacement should be considered at all sites. Just as important is the fact that standards vary by Region, indicating that interpretations must be tailored to soil type and climate. Regardless of the precise accuracy of these standards, this is a large step forward. The Forest Service has acted responsibly, progressively, and well in advance of the Montreal Process. Strengths of the Forest Service effort are that they provide a consistent and operationally practicable means for addressing the extent of significant soil degradation as called for by the Santiago Declaration. Shortcomings we see in such standards are that their true relationships with soil productivity have not been established, that subsoil conditions are not addressed, and that important soil processes are not integrated particularly well. However, standards will be updated as better findings surface.

Indicators of Integrated Properties and Processes

Most historical soil-site productivity research efforts in the USA have tried to relate soil properties to measures of productivity, but correlations usually were weak (Carmean, 1975). One weakness was that productivity generally was measured by tree site index, in itself an abstraction of site productivity. The earliest studies dealt with unregulated stands with uncertain histories. Soil variables tended to be static properties (horizon thickness, soil depth, drainage class, total or extractable nutrient content, and exchangeable cations) that represented form more than function. Usually, soil variables were evaluated additively through multiple regression with the premise that variables were linearly related to site index and were independent of each other. Generally, they are not. Interactions seldom were considered. Often, site index was estimated with minimal error for a stand, but soil conditions were determined from a single pit (producing independent variables with unknown error). Newer efforts have been more sophisticated; however, chemically extractable measures of soil nutrients, popular in agronomy, have rarely related well to forest productivity. The most useful variables invariably reflect some measure of water availability, rather than nutrient availability.

Another historical problem with soil–site studies is the preoccupation with surface conditions. While fine roots do concentrate in the forest floor and surface soil where organic matter decomposition and nutrient uptake predominate, larger roots can and do penetrate to great depths if water supplies can be reached and O_2 is adequate. Even genera widely thought to be shallow-rooted, such as *Picea*, root to depths of 2 to 6 m unless restricted by permafrost or a high water table, and drought-adapted genera such as *Eucalyptus, Juniperus,* and *Pinus* can root to depths of 20 m or more (Stone & Kalisz, 1991). Much deep root penetration is in saprolite at depths beyond the 200-cm limit used arbitrarily in U.S. soil mapping (Soil Survey Staff, 1996). The significance of deep rooting is ignored in most soil–site studies (Hammer et al., 1995; Richter et al., 1995; Stone & Kalisz, 1991) and probably reflects a predilection with nutrition, rather than water.

One of the earliest forestry efforts at combining quantitative soil properties into a useful index is that of Storie and Wieslander (1949). Soil depth, texture, permeability, chemistry, and drainage were multiplied to give an index of site quality. One value of this index was that site quality could be estimated for land units lacking suitable site index trees. Storie and Wieslander never meant for their soil rating index to be used to detect changes in soil quality, nor is it precise or sensitive enough to accomplish this, but their concept was a step forward. More recently, Jackson and Gifford (1974) proposed an integrative approach combining water balance, nutrient and temperature measures to estimate the growth potential of *Pinus radiata* in New Zealand. Henderson et al. (1990) urged that indices should be integrated over the entire soil profile. Using a concept pioneered by Kiniry et al. (1983) in agronomy and adapted to forestry by Gale (1987), they proposed that a series of soil physical and chemical sufficiency values correlated with root growth be assigned weights and combined in either an additive or multiplicative way to produce an index of potential root growth. Variables suitable for their region included soil pH, bulk density, aeration, and available water content. They suggested that measures of organic matter mineralization also might be useful, but they emphasized that the model should be simple enough that it could be applied to data collected routinely in soil surveys. Using soil electrical conductivity, available P and soil depth, Burger and Kelting (1998, this publication) applied this concept successfully to reclaimed mine spoils in the Appalachians.

RECOMMENDATIONS

We believe that soil quality standards meeting the spirit of the Santiago Agreement can be established for forestry in the USA. We also agree with Henderson et al. (1980) that soil quality indices must be simple enough to be practicable and must consider more than surface soil horizons. Methods requiring sophisticated and costly laboratory analyses *sensu* Dornan and Parker (1994) are impractical for extensive monitoring in forestry. Workable standards must have the following traits:

1. Indices must reflect physical, nutritional, and biological soil processes important to potential productivity.
2. Each index must integrate a variety of properties and processes.
3. Procedures must be practicable in an operational monitoring program and be applicable to a variety of soils.
4. Interpretations must be sensitive to the overriding conditions of climate.

Our recommendations cover three categories of soil quality indices: (i) a physical index that integrates soil density, structure, and moisture content; (ii) a nutritional index that integrates organic matter quality, content, and microbial activity; and (iii) a biological index that integrates the activity of soil organisms relative to physical and chemical soil properties. But such indices would have limited value without an adequate inventory of soil conditions. Therefore, our first recommendation is for an extensive and effective soil inventory.

Soil Inventory and Mapping

The first step in establishing a baseline for detecting changes in soil quality is a soil survey and mapping program that covers all forested regions and ownerships. For soil quality monitoring, the soil inventory will be used to ensure that present and future comparisons are made on the same population. The units should be based on stable soil properties and not vegetation or other attributes where changes may be confounded with changes in soil quality. The polygons should delineate functional soil units based on soil properties that have strong relationships to forest productivity, such as available soil water capacity and depth of rooting (Hammer et al., 1995; Henderson et al., 1990).

Other resource data can be overlaid, but the soil survey must stand on its own merits and not be confused with ecological classification systems which can contain bias based on interpretations of potential vegetation. Quality should at least meet the minimum standards of the National Cooperative Soil Survey (Soil Survey Staff, 1996). The scale for a mapping unit must be a compromise between that which is scientifically desirable and that which is economically feasible. Broad surveys must be useful to forest managers. We suggest that contrasting soils be mapped to a minimum of 4 ha, the minimum mapping unit used by the California State Cooperative Soil-Vegetation Survey (Gardner & Wieslander, 1957). This size is more in keeping with the minimal size of operational units to be managed in the foreseeable future. Areas designated for repeated soil quality monitoring, such as long-term research sites, should be mapped as intensively as economically feasible.

Historically, field descriptions were made from soil profiles in typical locales surveyed in the course of mapping. Some horizon properties were described in the field and others described following laboratory analyses of physical and chemical properties. These descriptions were used to describe typical profile values for each series and to classify the soil in a hierarchical system. This traditional system is appropriate in describing the properties affected by pedogenesis; however, we recommend that measurements emphasize properties that may be changed by management and whose function varies by climate. Furthermore, the geographically referenced profile sampling points also are potential reference points for repeated sampling in the future.

The two-stage process of mapping and describing the soils within each mapping unit understates the complexity of the soil on the landscape (Ponce-Hernandez, 1994), giving users a misleading picture of the variability involved. The uncertainty associated with grouping of natural bodies, locating boundaries, and portraying a gradual change with an abrupt line should be described with an error model or visual representation (Goodchild, 1994). The basis for this could be the variability around average values. Not generally available, these measures of natural variability are obtainable from existing data and should be described and available in published form. This establishes the normal range of variability within a functional soil unit, and provides the basis for describing changes in soil properties that relate to soil quality.

Indices of Physical Quality

Despite the popularity of bulk density as the standard for measuring physical soil impacts, we conclude that in itself, bulk density is not a particularly useful index of soil quality. Soil physical properties that influence productivity relate to supplies of O_2 and water, and the degree by which the soil matrix affects the proliferation of roots (da Silva et al., 1994). This was illustrated under controlled conditions by Simmons and Pope (1988; Fig. 3–3). Presently, the simplest and most practicable means of assessing soil physical properties associated with productivity is soil strength. Strength measures the force needed to overcome the resistance between soil particles to penetration by a small object such as a root tip. It increases with soil bulk density and decreases with soil moisture and organic matter content. Root growth declines exponentially with soil strength, being seriously restricted at 2 MPa (Taylor et al., 1966) and essentially ceasing at 3 MPa (Sands et al., 1979; Whalley et al., 1995).

Innovations in recording cone penetrometers now permit rapid and reproducible point estimates of profile strengths in the field, with minimal disturbance, to depths as great as 60 cm at depth intervals as small as 15 mm. Experience has shown that coefficients of variation average 6% when a soil is stratified into common disturbance classes (undisturbed, general disturbance, skid trails). Past problems of inconsistent entry rates largely have been overcome. Such technology is quite affordable and can be used to assess differences due to soil disturbance both spatially (Fig. 3–1) and temporally (Fig. 3–5).

Critics of soil strength have dismissed it as too sensitive to soil moisture to be a useful monitoring measure. We maintain that such sensitivity is the heart of its value. Figure 3–5 illustrates differences in soil strength as measured by penetrometer resistance between adjacent conifer stands growing on a sandy loam.

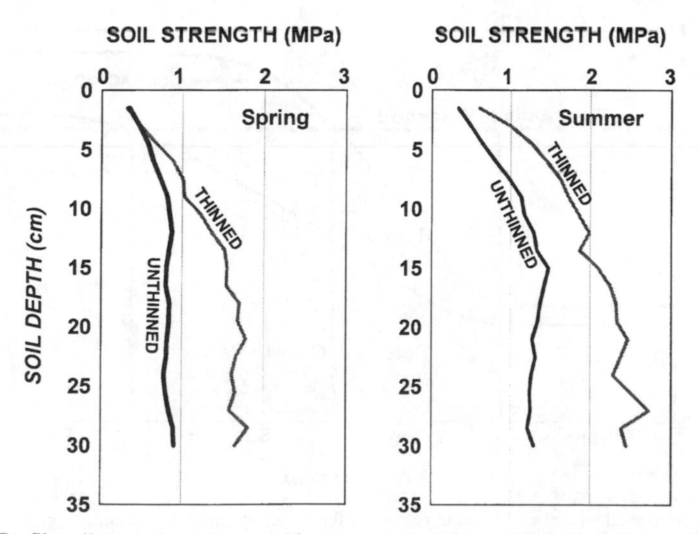

Fig. 3–5. Profile soil strengths as measured by cone penetrometer resistance on adjacent thinned and unthinned stands. Treatment differences increase as soils become drier in the summer (Powers, 1996, unpublished data).

One stand had been undisturbed since the early part of the century, while the other had received a mechanized thinning the previous year. Measurements taken during moist conditions in the spring, 1 yr after disturbance, show similar strengths between thinned and unthinned areas to a depth of 10 cm. Between 15 and 30 cm the differences widened to approximately 1 MPa, but soil beneath the thinned stand was approaching the 2 MPa threshold considered significant to plant growth. Strength always increases as soils dry, and differences had widened by summer. Soil strength in summer exceeded 2 MPa below 15 cm in the thinned stand and was approaching 3 MPa—the point at which root activity essentially ceases. Strength measurements taken nearby in skid roads created a century earlier during railroad logging still show detectable differences. Recording penetrometers also can detect conditions having limited extent but a major bearing on water infiltration and root growth (such as krotovinas and ped faces in shrink-swell clays), and can help in detecting their spatial arrangement.

Monitoring soil strength using the hand-held recording penetrometer is the simplest and most integrative means we know for inferring physical effects of disturbance on potential root activity to depths of up to 60 cm. The learning period is short and an experienced worker can obtain hundreds of reproducible samples a day. The sensitivity of strength to soil moisture as well as soil structure and density allows the charting of changes throughout a growing season (Fig. 3–6). The integral beneath the curve from the initiation of growth in the spring (when soil moisture is abundant) to the attainment of a determined root activity threshold (such as 3 MPa) should be an effective measure of the physical change in soil quality. Unfortunately, the penetrometer is ineffective in soils with high gravel content—a condition prevalent in many forest soils. We can offer no immediate remedy to this dilemma.

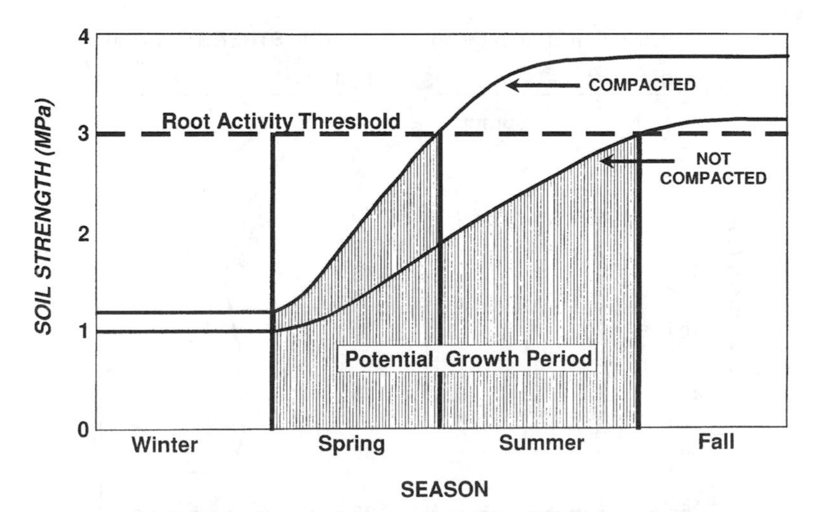

Fig. 3–6. Conceptual model of how compaction affects soil strength and potential growing season (shaded area). Regardless of soil texture, root activity (growth, water and nutrient uptake) is presumed to slow and essentially cease as soil strength within the rooting zone approaches 3 MPa (Powers & Avers, 1995).

da Silva et al. (1994) propose a more refined, physiologically based index of soil physical quality based on a concept introduced by Letey (1985). This index called least-limiting water range (LLWR) incorporates limitations due to aeration, soil strength, and water availability once rapid drainage has ceased. The objective is to define a soil's LLWR in cubic meters per cubic meter between limits set by aeration porosity (0.1 m^3 m^{-3}) at the wet end and soil strength (2 MPa) and drought ($\psi = -1.5$ MPa) at the dry end. This range reflects the conditions described in Fig. 3–3 and 3–6. Changes that lead to a narrowing of LLWR would be detrimental in that vegetation would be more vulnerable to drought or waterlogging. The index must be tailored for individual soils. In spite of the drawback that considerable modeling is needed under controlled conditions (favorable water content must be defined for important horizons between the limits of low aeration and high strength), the LLWR concept has considerable promise for application to forestry.

Indices of Nutrient Supply

Ideal nutrient supply indices would integrate the nature of the substrate, microbial activity, physical and chemical controls of decomposition, and weathering that ultimately affect the composition of the soil solution. But the immense variability of forest soils, the many assessment methods, and the cost of reliable laboratory work preclude intensive analyses for a range of nutrients in any operational program. Practicable tests must be simple to run, reproducible, integrative, sensitive to significant site changes, and correlated with the productive potential of soil. *Significant site changes* means that the test must be buffered against disturbance effects that are ephemeral. We believe that laboratory incubation of soil samples—particularly if sampling includes several soil layers—is the best single approach to indexing nutrient supply.

In 1980, Powers adapted the anaerobic incubation technique of Waring and Bremner (1964) to forest soils of California and Oregon and found that it correlated positively with site index and foliar N in *Pinus ponderosa* and response of several tree species to fertilization. Although it has mixed success in predicting fertilization response, the method has correlated well with tree growth and N uptake in such diverse forest sites as southeastern Australia (Adams & Attiwill, 1986), California (Powers, 1992), and Maine (Kraske & Fernandez, 1990). Anaerobically mineralized N measured in the laboratory under standard conditions has correlated remarkably well with rates of aerobic N mineralization in the field (Adams & Attiwill, 1986), particularly during periods when soil moisture is not limiting (Powers, 1990).

Anaerobically mineralized N has several advantages over other nutrient availability tests. It is an effective index of microbial biomass (Adams & Attiwill, 1986; Myrold, 1987). While microbial biomass alone is not a strong index of soil quality, it is recommended as a component of a more integrative index (Fauci & Dick, 1994). Advantages of anaerobic over aerobic incubations are that only a single species of N (ammonium) is analyzed and the need for careful balances between aeration and moisture is eliminated. Coefficients of variation for individual samples within apparently uniform soils average 45% (Powers, 1980), but is only one-half that when adjacent samples are composited. In the West, stress thresholds have been established at between 12 mg N kg^{-1} and 15 mg N kg^{-1}, depending on tree species (Powers, 1980, 1992). The sensitivity of anaerobic mineralization to substrate quality is shown by its strong negative correlation with soil C/N ratio and positive correlation with total organic C and N (Kraske & Fernandez, 1990; Powers, 1980). Anaerobically mineralized N and root distribution are parallel and decline with soil depth (Fig. 3–7). Describing mineralizable N profiles to a depth of up to 1 m incorporates the concept of Kiniry et al. (1983) as applied to forestry by Gale (1987) and Henderson et al. (1990), and would be more effective than monitoring any specific horizon or depth.

In advocating anaerobic incubation as a soil quality technique we are not implying that N availability is the only nutritional factor affecting soil productivity. Rather, we believe that microbial decomposition of soil organic matter is an important index that integrates many nutritional factors. It is particularly relevant in closed stands that depend on organic pathways of sustained nutrient supply. The capacity of anaerobic incubation to describe the mineralization potential of other organically supplied nutrients, particularly P, which owes much of its biological availability to microbial transformations (McGill & Cole, 1981; Gressel et al., 1996) merits further study.

Recently, Updegraff et al. (1994) described a more elaborate method for obtaining an anaerobic index of C, N, and P mineralization. Their method accounts for long-term changes in mineralization kinetics and identifies rank reversals that are not apparent in short-term incubations. Although it is elaborate, the Updegraff et al. (1994) method has an advantage in that one may model possible shifts in trace gas emissions to the atmosphere in waterlogged soils, which has obvious advantages in the study of wetlands. We believe that such tests, particularly if interpreted against regional standards or in *before and after* comparisons, offer the most promise for an effective nutrient index that blends substrate quality, quantity, and microbial activity.

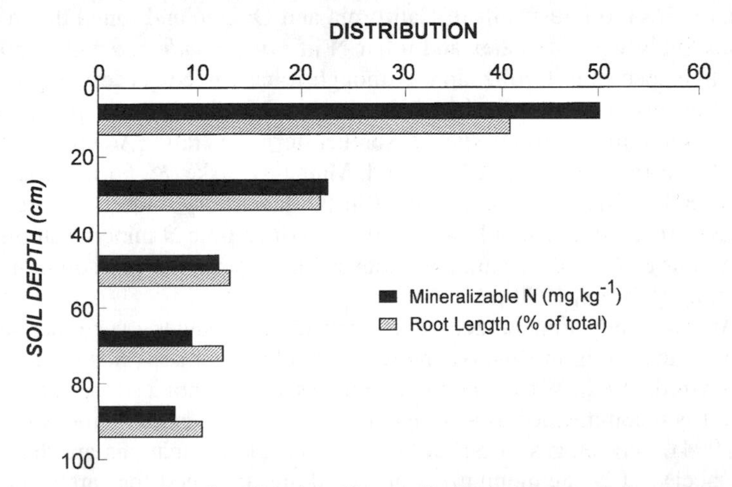

Fig. 3–7. Anaerobically mineralized N and conifer root distributions follow similar patterns (Powers, 1980). Root data are percentages of total root lengths to 1 m depth averaged for five species of conifers on a clay loam.

Indices of Soil Faunal Activity

The significance of soil fauna (specifically, invertebrates) to forest ecosystem processes is acknowledged far more widely in Europe than in the USA (Shaw et al., 1991). Yet, soil fauna affect essentially all of the soil properties and processes related to soil quality. Therefore, they must be considered in any soil quality monitoring program. Soil invertebrates are grouped by size into micro- (<0.1 mm), meso- (0.1 to 2 mm), and macrofauna (>2 mm)—classes that also reflect their function in soil processes (Linden et al., 1994). Microfauna are microbivores that regulate bacterial and fungal populations, thereby increasing nutrient turnover rates directly and soil aggregation indirectly. Mesofauna also feed on bacteria and fungi, and some consume feces of other soil fauna and plant detritus, thereby increasing fragmentation of organic residues and colonization by microbes. Larger mesofauna such as enchytraeids tunnel and increase soil porosity. In the process, materials passing as fecal pellets help form small soil aggregates. Soil macrofauna are the principal comminutors of organic residues from the soil surface into deeper regions of rooting. The digestive intermingling of mineral particles and organic residues increases surface area and generally increases substrate availability for microbes throughout the soil profile. Net rates of N mineralization (and presumably other nutrients, too) are substantially greater in soil aggregates and fecal material than in the whole soil (Sharpley & Syers, 1977; Sollins et al., 1984). Many macrofauna tunnel, thereby creating biopores that enhance soil aeration, water infiltration, and root growth. The multiple roles of soil fauna in forests have been reviewed by Shaw et al. (1991).

Some soil invertebrates seem particularly sensitive to management-caused disturbances. Hoekstra et al. (1995) found that numbers of predaceous litter arthropods were reduced for up to 15 yr following selective harvesting in *Sequoia sempervirens*, although other functional guilds were not affected. Logically, soil invertebrates have a place as indicators of forest soil quality and sustainable productivity. But collecting, identifying and quantifying soil fauna are complex and specialized tasks (Moldenke, 1994). For example, earthworm biomass can triple within a month (Karlen et al., 1997). Therefore, we do not advocate detailed inventories of soil invertebrates as an element of soil quality monitoring. Instead, attention should turn to a narrower group of macrofauna. Because of their sensitivity to disturbance and their position in the food chain, predaceous litter spiders have been proposed as indicators of biodiversity (McIver et al., 1990). But the significance of predaceous species in soil ecosystem processes is speculative, at best. We advocate function over form, and propose the macroshredders such as anecic earthworms, millipedes, and ants as a particularly relevant group. These are the most conspicuous soil invertebrates and comprise the group with the greatest potential as biological indicators of important soil processes.

The concept of bioindicators has its critics (Linden et al., 1994). Tolerant organisms can exist in many environments and classifying soil fauna by their tolerance or sensitivity to disturbance is largely subjective. Further, species distribution is determined by many environmental factors other than disturbance. Soil fauna also have temporal patterns of behavior. Earthworms apparently absent during a dry season suddenly may surface during a rain. Another criticism of indicator species is that they become charismatic. They convey a keystone impression of a critical, cause-and-effect role in soil function. In fact, the role of invertebrate fauna is more that of a catalyst of processes that would proceed without them, albeit at a slower rate. Because of this, we conclude that the most promising approach for operational monitoring lies not in monitoring macroshredders directly—but rather, the products of their activity. Biopores, soil aggregates, and fecal deposits are recognized readily, are not dependent on seasonal effects, and will degrade if activity declines for an extended period. Thus, the physical products of soil invertebrates—particularly, the macroshredders—have far greater value than the presence or apparent absence of the specific organisms. We consider this a realm of forest ecosystem relationships and research that is ripe for continuing work.

SUMMARY

The Santiago Declaration calls for the world's forested nations to adopt standards for the conservation and maintenance of soil and water resources in support of sustainable forestry. Among these are indicators of significant soil erosion, diminished organic matter and other chemical attributes, and degradation of physical properties. Paralleling this is the effort of the Soil Science Society of America to develop soil quality criteria for a sustainable environment. We endorse these concepts as guides to sustainable forest management in the USA. However, we caution that the sweep of environmental conditions and management strategies that bear on forests of the nation makes intensive sampling and detailed analyses impractical in most cases. In fact, we doubt that an extensive list of static indicators has any particular relevance for ecosystems with changing plant communities and the complexity of plant–soil interactions affecting long-lived vegetation. The USDA Forest Service recognizes this. Their soil quality interpretations vary with soil type and climate. Region by region, threshold standards are being set for detecting detrimental changes in soil quality. These are seen as interim guides for National Forests because they await validation through the LTSP program.

As the concept of soil quality evolves, so do standards for effective monitoring. For now, we propose that a smaller set of indicators be adopted regionally that integrate to the fullest extent practicable the soil factors known to influence forest productivity. These should be the simplest possible indices of key physical, nutritional, and biological soil properties and processes. We propose soil strength, nutrient release through anaerobic incubation of organic matter, and signs of soil invertebrate activity as the most operationally feasible and useful indices at the present. Such indices should not be used in an absolute sense (a specific standard of quality). Instead, they can be used to establish baseline conditions on a forested site. The same site would be sampled again to monitor changes in these indices, but not until initial perturbation effects have subsided.

REFERENCES

Adams, M.A., and P.M. Attiwill. 1986. Nutrient cycling and nitrogen mineralization in eucalypt forests of south-eastern Australia: II. Indices of nitrogen mineralization. Plant Soil 92:341–362.

Alban, D.H., G.E. Host, J.D. Elioff, and J.A. Shadis. 1994. Soil and vegetation response to soil compaction and forest floor removal after aspen harvesting. Res. Pap. NC-315. USDA Forest Serv. North Central For. Exp. Stn., St. Paul, MN.

Allen, H.L., L.A. Morris, and T.R. Wentworth. 1991. Productivity comparisons between successive loblolly pine rotations in the North Carolina Piedmont. p. 125–136. *In* W.J. Dyck and C.A. Mees (ed.) Long-term field trials to assess environmental impacts of harvesting. For. Res. Inst. New Zealand Bull. 161. Ministry of For., For. Res. Inst., New Zealand.

American Forest and Paper Association. 1994. Sustainable forestry principles and implementation guidelines. AFPA,Washington, DC.

Bennett, H.H. 1939. Soil conservation. McGraw-Hill, New York.

Berg, S., and R. Olszewski. 1995. Certification and labeling. A forest industry perspective. J. For. 93:30–31.

Burger, J.A., and D.L. Kelting. 1998. Soil quality monitoring for assessing sustianable forest management. p. 17–52. *In* The contribution of soil science to the development of and implementation of criteria and indicators of sustainable forest management. SSSA Spec. Publ. 53. SSSA, Madison, WI (this publication).

Burger, J.A., and R.A. Kluender. 1982. Site preparation: Piedmont. p. 58–74. *In* R.C. Kellison and S.A. Gingrich (ed.) Symposium on the Loblolly Pine Ecosystem (East Region). USDA For. Serv. and North Carolina State Univ., Raleigh, NC.

Busse, M.D., P.H. Cochran, and J.W. Barrett. 1996. Changes in ponderosa pine site productivity following removal of understory vegetation. Soil Sci. Soc. Am. J. 60:1614–1621.

Carmean, W.H. 1975. Forest site quality evaluation in the United States. Adv. Agron. 27:209–269.

Compton, J.E., and D.W. Cole. 1991. Impact of harvest intensity on growth and nutrition of successive rotations of Douglas-fir. p. 151–161. *In* W.J. Dyck and C.A. Mees (ed.) Long-term field trials to assess environmental impacts of harvesting. For. Res. Inst. New Zealand Bull. 161. Ministry of For., For. Res. Inst., New Zealand.

da Silva, A.P., B.D. Kay, and E. Perfect. 1994. Characterization of the least limiting water range of soils. Soil Sci. Soc. Am. J. 58:1775–1781.

Doran, J.W., D.C. Coleman, D.F. Bezdicek, and B.A. Stewart (ed.). 1994. Defining soil quality for a sustainable environment. SSSA Spec. Publ. 35. SSSA, Madison, WI.

Doran, J.W., and T.B. Parkin. 1994. Defining and assessing soil quality. p. 3–21. *In* J.W. Doran et al. (ed.) Defining soil quality for a sustainable environment. SSSA Spec. Publ. 35. SSSA, Madison, WI.

Fauci, F., and R.P. Dick. 1994. Microbial biomass as an indicator of soil quality: Effects of long-term management and recent soil amendments. p. 229–234. *In* J.W. Doran et al. (ed.) Defining soil quality for a sustainable environment. SSSA Spec. Publ. 35. SSSA, Madison, WI.

Froehlich, H.A., and D.H. McNabb. 1984. Minimizing soil compaction in Pacific Northwest forests. p. 159–192. *In* E.L. Stone (ed.) Forest soils and treatment impacts. Proc. 6th North Am. Forest Soils Conf. Univ. of Tennessee, Dep. of Forestry, Wildlife and Fisheries, Knoxville.

Fleming, R.L., T.A. Black, and N.R. Eldridge. 1994. Effects of site preparation on root zone water regimes in high-elevation forest clearcuts. For. Ecol. Manage. 68:173–188.

Gale, M.R. 1987. A forest productivity index model based on soil- and root-distributional characteristics. Ph.D. diss. Univ. of Minnesota, St. Paul.

Gardner, R.A., and A.E. Wieslander. 1957. The soil vegetation survey in California. Proc. Soil Sci. Soc. Am. 2:103–105.

Goodchild, M.F. 1994. GIS error models and visualization techniques for spatial variability of soils. p. 683–698. *In* 15th World Congr. of Soil Sci. Mexican Soc. Soil Sci., Mexico City.

Grable, A.R., and E.G. Siemer. 1968. Effects of bulk density, aggregate size, and soil water suction on oxygen diffusion, redox potentials and elongation of corn roots. Soil Sci. Soc. Am. Proc. 32:180–186.

Gressel, N., J.G. McColl, C.M. Preston, R.H. Newman, and R.F. Powers. 1996. Linkages between phosphorus transformations and carbon decomposition in a forest soil. Biogeochemistry 33:97–123.

Hammer, R.D., G.S. Henderson, R.P. Udawatta, and D.K. Brandt. 1995. Soil organic carbon in the Missouri forest-prairie ecotone. p. 201–231. *In* W.W. McFee and J.M. Kelly (ed.) Carbon forms and functions in forest soils. Proc. 8th North Am. Forest Soils Conf., Soil Sci. Soc. Am., Madison, WI.

Harrison, R.B., S.P. Gessel, D. Zabowski, C.L. Henry, D. Xue, D.W. Cole, and J.E. Compton. 1996. Mechanisms of negative impacts of three forest treatments on nutrient availability. Soil Sci. Soc. Am. J. 60:1622–1628.

Henderson, G.S., R.D. Hammer, and D.F. Grigal. 1990. Can measurable soil properties be integrated into a framework for characterizing forest productivity? p. 137–154. *In* S.P. Gessel et al. (ed.) Sustained productivity of forest soils. Proc. 7th North Am. Forest Soils Conf., Univ. British Columbia, Faculty of Forestry, Vancouver, BC.

Hoekstra, J.M., R.T. Bell, A.E. Launer, and D.D. Murphy. 1995. Soil arthropod abundance in coast redwood forest: effect of selective timber harvest. Environ. Entomol. 24:246–252.

Jackson, D.S., and H.H. Gifford. 1974. Environmental variables influencing the increment of radiata pine. NZ J. For. Sci. 4:3–26.

Journal of Forestry. 1995. The certified forest: what makes it green? J. For. 93:1–41.

Karlen, D.L., M.J. Mausbach, J.W. Doran, R.G. Cline, R.F. Harris, and G.E. Schuman. 1997. Soil quality: A concept, definition, and framework for evaluation. Soil Sci. Soc. Am. J. 61:4–10.

Kiniry, L.N., C.L. Scrivner, and M.E. Keener. 1983. A soil productivity index based upon predicted water depletion and root growth. Missouri Agric. Stn. Res. Bull. 1051. Columbia, MO.

Kraske, C.R., and I.J. Fernandez. 1990. Conifer seedling growth response to soil type and selected nitrogen availability indices. Soil Sci. Soc. Am. J. 54:246–251.

Larson, W.E., and F.J. Pierce. 1994. The dynamics of soil quality as a measure of sustainable management. p. 37–51. *In* J.W. Doran et al. (ed.) Defining soil quality for a sustainable environment. SSSA Spec. Publ. 35. SSSA, Madison, WI.

Letey, J. 1985. Relationship between soil physical properties and crop production. Adv. Soil Sci. 1:277–294.

Linden, D.R., P.F. Hendrix, D.C. Coleman, and P.C.J. van Vliet. 1994. Faunal indicators of soil quality. p. 91–101. *In* J.W. Doran et al. (ed.) Defining soil quality for a sustainable environment. SSSA Spec. Publ. 35. SSSA, Madison, WI.

McGill, W.B., and C.V. Cole. 1981. Comparative aspects of cycling of organic C, N, S and P through soil organic matter. Geoderma 26:267–286.

McIver, J.D., A.R. Moldenke, and G.L. Parsons. 1990. Litter spiders as bio-indicators of recovery after clearcutting in a western coniferous forest. Northwest Environ. J. 6:410–412.

Messier, C. 1993. Factors limiting early growth of western redcedar, western hemlock and Sitka spruce seedlings on ericaceous-dominated clearcut sites in coastal British Columbia. For. Ecol. Manage. 60:181–206.

Miller, R.E., W. Scott, and J.W. Hazard. 1996. Soil compaction and conifer growth after tractor yarding at three coastal Washington locations. Can. J. For. Res. 26:225–236.

Moen, J.E.T. 1988. Soil protection in the Netherlands. p. 1495–1503. *In* K. Wolf et al. (ed.) Contaminated soil. Kluwer Acad. Publ., Amsterdam.

Moldenke, A.R. 1994. Arthropods. p. 517–542. *In* Methods of soil analysis. Part 2. Microbiological and Biochemical Properties. SSSA Book Ser. 5. SSSA, Madison, WI.

Morris, L.A., and R.E. Miller. 1994. Evidence for long-term productivity change as provided by field trials. p. 41–80. *In* W.J. Dyck et al. (ed.) Impacts of forest harvesting on long-term site productivity. Chapman & Hall, London.

Myrold, D.D. 1987. Relationship between microbial biomass and a nitrogen availability index. Soil Sci. Soc. Am. J. 51:1047–1049.

Nambiar, E.K.S. 1996. Sustained productivity of forests is a continuing challenge to soil science. Soil Sci. Soc. Am. J. 60:1629–1642.

Ponce-Hernandez, R. 1994. Improving the representation of soil spatial variability in geographical information systems: A paradigm shift and its implications. p. 780–801. *In* 15th World Congr. Soil Sci.

Powers, R.F. 1980. Mineralizable soil nitrogen as an index of nitrogen availability to forest trees. Soil Sci. Soc. Am. 44:1314–1320.

Powers, R.F. 1990. Nitrogen mineralization along an altitudinal gradient: interactions of soil temperature, moisture, and substrate quality. For. Ecol. Manage.30:19–29.

Powers, R.F. 1992. Fertilization response of subalpine *Abies* forests in California. p. 114–126. *In* H.N. Chappell et al. (ed.) Forest fertilization: Sustaining and improving nutrition and growth of western forests. Inst. Forest Resour. Contrib. 73. Univ. of Washington, Seattle.

Powers, R.F. 1999. On the sustainable productivity of planted forests. New Forests (In press).

Powers, R.F., D.H. Alban, R.E. Miller, A.E. Tiarks, C.G. Wells, P.E. Avers, R.G. Cline, R.O. Fitzgerald, and N.S. Loftus, Jr. 1990. Sustaining site productivity in North American forests: Problems and prospects. p. 49–79. *In* S.P. Gessel et al. (ed.) Sustained productivity of forest soils. Proc. 7th North Am. Forest Soils Conf. Univ. British Columbia, Faculty of Forestry, Vancouver, BC.

Powers, R.F., and P.E. Avers. 1995. Sustaining forest productivity through soil quality standards: A coordinated U.S. effort. p. 147–190. *In* C.B. Powter et al. (ed.) Environmental soil science:

Anthropogenic chemicals and soil quality criteria. Canadian Soc. Soil Sci., Brandon, Manitoba.

Powers, R.F., and G.T. Ferrell. 1996. Moisture, nutrient, and insect constraints on plantation growth: The "Garden of Eden" study. NZ J. For. Sci. 26 (1/2):126–144.

Powers, R.F., A.E. Tiarks, J.A. Burger, and M.C. Carter. 1996. Sustaining the productivity of planted forests. p. 97–134. *In* M.C. Carter (ed.) Growing trees in a greener world: Industrial forestry in the 21st century. Proc. 35th LSU Symp., School of For., Wildlife and Fisheries, Louisiana State Univ., Baton Rouge.

Reicosky, D.C., W.B. Voorhees, and J.K. Radke. 1981. Unsaturated water flow through a simulated wheel track. Soil Sci. Soc. Am. J. 45:3–8.

Richter, D.D., D. Markewitz, J.K. Dunsomb, P.R. Heine, C.G. Wells, A. Stuanes, H.L. Allen, B. Urrego, K. Harrison, and G. Bonani. 1995. Carbon cycling in a loblolly pine forest: Implications for the missing carbon sink and for the concept of soil. p. 233–251. *In* W.W. McFee and J.M. Kelly (ed.) Carbon forms and functions in forest soils. Proc. 8th North Am. For. Soils Conf. Soil Sci. Soc. Am., Madison, WI.

Sands, R., E.L. Greacen, and C.J. Girard. 1979. Compaction of sandy soils in radiata pine forests: I. A penetrometer study. Aust. J. Soil Res. 17:101–113.

Senyk, J., and D. Craigdallie. 1997. Effects of harvesting methods on soil properties and forest productivity in interior British Columbia. Canadian Forest Serv. Info. Rep. BC-X-365. Pacific Science Ctr., Victoria, BC.

Sharpley, A.N., and J.K Syers. 1977. Seasonal variation in casting activity and in the amount and release to solution of phosphorus forms in earthworm casts. Soil Biol. Biochem. 9:227–231.

Shaw, C.H., H. Lundqvist, A. Moldenke, and J.R. Boyle. 1991. The Relationships of soil fauna to long-term forest productivity in temperate and boreal ecosystems: processes and research strategies. p. 39–77. *In* W.J. Dyck and C.A. Mees (ed.) Long-term field trials to assess environmental impacts of harvesting. For. Res. Inst. New Zealand Bull. 161. Ministry of For., For. Res. Inst., New Zealand.

Sheffield, R.M., and N.D. Cost. 1986. Evidence of pine growth loss in Forest Service inventory data. p. 74–85. *In* Atmospheric deposition and forest productivity. Proc. 4th Regional Technical Conf., Appalachian Soc. Am. Foresters, Raleigh, NC.

Sheffield, R.M., N.D. Cost, W.A. Bechtold, and J.P. McClure. 1985. Pine growth reductions in the southeast. Resource Bull. SE-83. USDA For. Serv. Southeast. Res. Stn., Asheville, NC.

Simmons, G.L., and P.E. Pope. 1988. Influence of soil water potential and mycorrhizal colonization on root growth of yellow poplar and sweet gum seedlings grown in compacted soil. Can. J. For. Res. 18:1392–1396.

Smethurst, P.J., and E.K.S. Nambiar. 1990. Effects of contrasting silvicultural practices on nitrogen supply to young radiata pine. p. 85–96. *In* W.J. Dyck and C.A. Mees (ed.) Impact of intensive harvesting on forest site productivity. IEA/BE T6/A6 Rep. 2. For. Res. Inst. Bull. 159. For. Res. Inst., Rotorua, NZ.

Soil Survey Staff. 1996. Keys to soil taxonomy. 7th ed. USDA Natural Resource Conserv. Serv., Washington, DC.

Sollins, P., G. Spycher, and C.A. Glassman. 1984. Net nitrogen mineralization from light- and heavy-fraction forest soil organic matter. Soil Biol. Biochem. 16:31–37.

Stone, E.L. 1975. Soil and man's use of forest land. p. 1–9. *In* B. Bernier and C.H. Wingate (ed.) Forest soils and forest land management. Proc. 4th North Am. Forest Soils Conf. Laval Univ. Press, Quebec.

Stone, E.L., and P.J. Kalisz. 1991. On the maximum extent of tree roots. For. Ecol. Manage. 46:59–102.

Storie, R.E., and A.E. Wieslander. 1949. Rating soils for timber sites. Soil Sci. Soc. Am. Proc. 13:499–509.

Stransky, J.J., J.J. Roese, and K.G. Watterston. 1985. Soil properties and pine growth affected by site preparation after clearcutting. Southern J. Appl. For. 9:40–43.

Taylor, H.M., G.M. Robertson, and J.J. Parker, Jr. 1966. Soil strength-root penetration relations for medium to coarse-textured soil materials. Soil Sci. 102:18–22.

Tiarks, A.E., and J.D. Haywood. 1996. Effects of site preparation and fertilization on growth of slash pine over two rotations. Soil Sci. Soc. Am. J. 60:1654–1663.

Updegraff, K., S.D. Brigham, J. Pastor, and C.A. Johnson. 1994. A method to determine long-term anaerobic carbon and nutrient mineralization in soils. p. 209–219. *In* J.W. Doran et al. (ed.) Defining soil quality for a sustainable environment. SSSA Spec. Publ. 35. SSSA, Madison, WI.

USDA Forest Service. 1983. The principal laws relating to Forest Service activities. USDA Agric. Handb. 453. USDA For. Serv., Washington, DC.

Vanclay, J.K. 1996. Assessing the sustainability of timber harvests from natural forests: Limitations of indices based on successive harvests. J. Sustain. For. 3(4):47–58.

Van Cleve, K., W.C. Oechel, and J.L. Hom. 1990. Response of black spruce (*Picea mariana*) ecosystems to soil temperature modification in interior Alaska. Can. J. For. Res. 20:1530–1535.

Vomocil, J.A., and W.J. Flocker. 1961. Effect of soil compaction on storage and movement of soil air and water. Trans. ASAE 4:242–245.

Warkentin, B.P. 1995. The changing concept of soil quality. J. Soil Water Conserv. 50:226–228.

Waring, S.A., and J.M. Bremner. 1964. Ammonium production in soil under waterlogged conditions. Nature (London) 201:951–952.

Whalley, W.R., E. Dumitru, and A.R. Dexter. 1995. Biological effects of soil compaction. Soil Tillage Res. 35:53–68.

Whitney, M. 1909. Soils of the United States, based upon the work of the Bureau of Soils to January 1, 1909. Bureau Soils Bull. 55. USDA, Washington, DC.

4 Can a Plot-Based Forest Health Monitoring System Contribute to Assessment of Soil Pollution Indicators in Canada?

Ian K. Morrison

Canadian Forest Service
Great Lakes Forestry Centre
Sault Ste. Marie, Ontario, Canada

As both forest and societal conditions vary from country to country, provisions of the Santiago agreement allow for national criteria and indicator (C&I) adaptations (Canadian Forest Service, 1995). In Canada, the federal and provincial governments, through the Canadian Council of Forest Ministers (CCFM) further defined indicators for Canadian application (Canadian Council of Forest Ministers, 1995). These CCFM indicators, while in some instances stated differently from those of the Montreal Process, are designed to be compatible with the international processes but provide more detail and precision on particular values of importance in Canada. Some 83 indicators are specified under the CCFM Process, compared with 67 under the Montreal Process. Other indicator cross-linkages also are apparent. The Canadian Standards Association (CSA) Sustainable Forest Management System standards, for example, are designed to be compatible with standards of the International Organization for Standardization and draw heavily on the CCFM indicators as well (Canadian Standards Association, 1996). For consistency, however, the international (i.e., Montreal Process) compilation is followed here.

Two of the 67 Montreal Process indicators are concerned with effects of air pollution on forests and/or forest soils, that is (i) area and percentage of forest land subjected to levels of specific air pollutants (e.g., sulphates, nitrate, ozone) or ultraviolet B that may cause negative impacts on forest ecosystems (Indicator 3b), and (ii) area and percentage of forest land experiencing an accumulation of persistent toxic substances (Indicator 4h). In Canada in particular, with a long history of air pollution damage to forests and, presumably, forest soils as well, mainly from metal-ore smelting, and with more-recent concern over possible effects of widespread, low-level pollutants, considerable interest has arisen in pollution monitoring. It is indicators of soil pollution that are examined in this chapter.

The Contribution of Soil Science to the Development of and Implementation of Criteria and Indicators of Sustainable Forest Management. SSSA Special Publication no. 53.

POINT-SOURCE AIR POLLUTANTS

Historically, air pollution injury to forests and forest soils in Canada was associated with a small number of significant, virtually-unchecked point-source emitters, chiefly metal-ore smelters. Damages, while often highly-visible, were tolerated as a price of economic progress. One case, which gained early notoriety, centered around a large Pb–Zn–Ag-ore smelter operated by the Consolidated Mining and Smelting Company of Canada at Trail, British Columbia, SO_2 fumes from which drifted into the state of Washington and damaged forests there (National Research Council of Canada. Associate Committee on Trail Smelter Smoke, 1939). The resulting mitigative actions, while a model of control and of recovery of by-products from waste, were probably compelled as much by a need to prevent damage to international relations as damage to forest stands and, with no national upwelling of environmental consciousness, decades passed before similar controls were initiated at other sources. Other well-documented instances of direct and sometimes striking injury to forests and/or forest soils include elemental S depositions around sour-gas processing plants in western Alberta (Baker, 1977; Addison et al., 1984), SO_2 and heavy metal damage to forests and soils around a Cu–Zn smelter in Flin Flon, Manitoba (Hogan & Wotton, 1984), a Ni smelter in Thompson, Manitoba (Blauel & Hocking, 1974), an Fe-ore sintering plant in Wawa, Ontario (Gordon & Gorham, 1963) and a Cu smelter in Murdochville, Quebec (Leblanc et al., 1974; Robitaille et al., 1977), and F injury to vegetation around an elemental P-producing factory in Long Harbour, Newfoundland (Sidhu & Roberts, 1976; Sidhu & Staniforth, 1986). One extreme example of smelter smoke injury stands apart, however, because of its magnitude. This centers around a complex of Ni–Cu ore smelters and refineries near Sudbury, Ontario, with at least one source (Dreisinger, 1970) estimating the area subject to at least one potentially injurious fumigation per growing season during the period before significant emission control (which commenced in the late 1960s) to be in excess of 5000 km^2. Both SO_2 fumes and heavy metal fall-out, then, have long been implicated as causes of forest damage in the Sudbury area (White & Linzon, 1950; Linzon, 1958; Gorham & Gordon, 1960; Hutchinson & Whitby, 1974; Whitby & Hutchinson, 1974; Freedman & Hutchinson, 1980; Amiro & Courtin, 1981; Lozano & Morrison, 1981).

Instances of extreme fume damage to soils, especially those involving multiple pollutants, represent only a very small percentage of the total forest area of Canada, so might otherwise be thought to be of only minor concern, save that they are important establishing a frame of reference for soil change. By way of illustration, the pollutants of concern in the Sudbury example, above, are SO_2 gas, and Ni^{2+} and Cu^{2+} particulate fall-out. In Table 4–1, properties of soils at various distances from the source area, but which were presumably originally similar, are given. The severely damaged soil of Table 4–1, 3 km downwind of the easternmost source, was devoid of vegetation except for a very-scattered remnant of depauperate trees. The site of moderate damage was 16 km downwind, whereas the control site was 24 km upwind of the westernmost source. As expected, soil S, measured as soluble SO_4^{2-}, increased as did total Ni^{2+} and total Cu^{2+} concentration with proximity to the source area. These changes occurred in both the sur-

Table 4–1. Properties of surface soil and subsoil of three soils near Sudbury, ON (adapted from Lozano & Morrison, 1981).

		Surface			Subsoil		
		Control	Moderate damage	Severe damage	Control	Moderate damage	Severe damage
OM	%	48.79a†	55.61a	--	6.63c	6.12c	3.29d
pH		4.26a	3.86b	--	4.59c	4.57c	4.52c
Total N	%	1.22a	1.26a	--	0.19c	0.16c	0.09d
Available P	$\mu g\ g^{-1}$	72a	48b	--	33c	12d	18e
Exchangeable K	$cmol\ kg^{-1}$	1.19a	0.84b	--	0.19c	0.10d	0.08d
Exchangeable Ca	$cmol\ kg^{-1}$	11.16a	3.81b	--	1.13c	0.40d	0.46d
Exchangeable Mg	$cmol\ kg^{-1}$	2.98a	0.86b	--	0.32c	0.10d	0.13d
Exchangeable Na	$cmol\ kg^{-1}$	0.14a	0.17b	--	0.07c	0.05d	0.04e
Exchangeable Al	$cmol\ kg^{-1}$	0.22a	0.32b	--	0.28c	0.17d	0.08e
Exchangeable H	$cmol\ kg^{-1}$	45.02a	57.13b	--	12.15c	6.71d	5.76d
Cation-exchange capacity	$cmol\ kg^{-1}$	60.49a	62.81a	--	13.86c	7.56d	6.47d
Base saturation	%	25.57a	9.04b	--	12.33c	8.83d	10.97c
Soluble SO_4^{2-}	$\mu g\ g^{-1}$	290a	645b	--	109c	222c	281d
Total Fe	$\mu g\ g^{-1}$	21 400a	27 800b	--	34 250c	33 000c	19 625d
Total Mn	$\mu g\ g^{-1}$	958a	358b	--	431c	386c	230d
Total Zn	$\mu g\ g^{-1}$	152a	111b	--	83c	64c	48d
Total Cu	$\mu g\ g^{-1}$	23a	75b	--	56c	62c	119d
Total Ni	$\mu g\ g^{-1}$	498a	1290b	--	114c	176c	252d
Total Al	$\mu g\ g^{-1}$	32 300a	28 200b	--	65 000c	57 800d	54 188e

† Differences in letters in rows (surface and subsoil separately) indicate significance at P=0.05.

face material (the soil of the severely damaged area lacked a recognizable surface organic layer, hence are not reported) and in the underlying mineral soil. As expected also, soil bases, measured as exchangeable K^+, Ca^{2+}, Mg^{2+}, and Na^+ generally decreased along the same gradient. Interestingly, these latter collateral changes, however, obvious to soil scientists, would not necessarily have been captured under either of the C&I pollution-related indicators as presently stated though, in this particular case, some sense of impact might have been ascertained in an abstruse way under Indicator 3c (diminished biological components indicative of changes in fundamental ecological processes) or Indicator 4d (significantly diminished soil organic matter and/or changes in other soil chemical properties).

LONG-RANGE TRANSPORTED AIR POLLUTANTS

During the 1970s, concern grew both in Europe and North America on possible effects of *acid deposition* or *acid rain,* i.e., natural precipitation augmented with strong mineral acids from the oxidation products of S and N, on natural environments (Tamm & Cowling, 1977; Glass et al., 1979, 1980). At first, in North America at least, depositions of manmade acids were considered to be problems mainly for aquatic ecosystems, and most research was focused in that direction. Research on forests or forest soils was more limited (Morrison, 1984). By the early 1980s, however, concerns arose in Europe about a new-type forest decline (neuartige Waldschäden) and fears were expressed of irreparable damages to forests, especially in central Europe (Ulrich et al., 1980). This was rapidly followed, then, by reports of new declines elsewhere: red spruce (*Picea rubens* Sarg.) in the Appalachian region of USA (Vogelmann, 1982; Tomlinson, 1983); white birch (*Betula papyrifera* Marsh.; Magasi, 1989) and sugar maple (*Acer saccharum* Marsh.) in eastern Canada (e.g., McLaughlin et al., 1985; Carrier & Gagnon, 1985; Carrier, 1986). In the latter case especially, well-defined patterns of injury were difficult to recognize, thus no straightforward, wide scale, acid deposition-oriented forest damage survey was undertaken as in Europe (United Nations-Economic Commission for Europe, and European Commission, 1992, 1993, 1994, 1995, 1996; European Commission, 1989, 1990, 1991a,b). Rather, multiple-stress hypotheses, with acid precipitation as a *contributing* but not the *sole* cause, were generally favored (Hendershot & Jones, 1989; Bertrand et al., 1994). In fact, in retrospect, during the course of a large USA–Canada study of sugar maple decline, no relationship was indeed found between levels of acidic deposition and tree crown condition, though it was acknowledged that impacts of pollutants could not be eliminated given that factors such as soil properties could synergize with deposited levels. Other stressors including drought, insect defoliation and soil properties, it was concluded, probably had greater impact (Lachance et al., 1995).

There was in place in Canada during this period as well, significant capacity for insect and disease survey, focused mainly on potentially damaging insect and diseases, and some capacity for other aspects of ecosystem research. Responding to a political imperative, and proposing to detect early signs of air pollution injury against the background of other stressors, a new network of biomonitoring plots was established in 1984 to supplement other more-extensive survey methods and plot networks. In keeping with the times, this was called the Acid Rain National Early Warning System (ARNEWS). The overall objectives of the ARNEWS, as currently stated (D'Eon et al., 1994), are to: (i) detect, clearly and accurately, damage to forest tree and soils caused by air pollutants or to identify damage sustained by Canada's forests (trees and soils) that is not attributable to natural causes or management practices; and (ii) monitor vegetation and soils to detect long-term changes attributable to air pollutants in representative forest ecosystems.

ARNEWS assessments attempt to detect subtle growth reductions or subtle damages, such as those (if any) related to regional air pollutants, in a two-step process: first, by eliminating other possible causes; then, if there are injuries that cannot be attributed to some known agent, researching these more thoroughly (Addison, 1989).

The ARNEWS is a plot-based biomonitoring system (D'Eon & Power, 1989). ARNEWS plots (Fig. 4–1) appear as 10 × 40 m permanent plots, i.e., 0.04-ha, with subplots. There are currently 150 ARNEWS locations across Canada, positioned to assess long-range transported air pollutant as opposed to short-range smoke damage effects (Fig. 4–2). In the field, the ARNEWS program functions as a series of assessments, conducted at various intervals (Table 4–2). Amongst these are (i) assessments several times per growing season for various abiotic symptoms as well as signs of insect or disease, (ii) annual assessment of tree condition and tree mortality, (iii) biennial assessments of regeneration and ground vegetation and, every 5 yr, assessments of (iv) tree growth, (v) tree foliage chemical composition, and (vi) soil chemical properties (Magasi, 1988; D'Eon et al., 1994). Results from annual and biennial assessments have been regularly summarized in a series of reports (Hall, 1991, 1993; Hall & Pendrel, 1992).

Of direct interest in the present C&I context is the monitoring of soil chemical properties conducted at each ARNEWS location at 5-yr intervals (Magasi, 1988; D'Eon et al., 1994) and the relevance of this to soil indicator assessment. First, it should be noted that field investigations of soils subjected to regional air pollution present a special problem for, unlike soil sample for other purposes where controls can be arranged or nearby untrammeled areas can usually be found and used for reference purposes, the widespread deposition patterns associated with long-range transported materials tends to preclude such approaches. The objective of the soil sampling and analysis program of the ARNEWS, then, is to assess change in a variety of mainly chemical soil properties over time.

Many of the plots are located in areas where conventional soil survey information is not available. Thus, a necessary first step is a baseline characterization of soil at each plot at the time of plot establishment wherein the soil described in detail, classified according to the Canadian System of Soil Classification (Canada Soil Survey Committee, 1978), and sampled and analyzed for a suite of parameters (D'Eon et al., 1994). Further, as natural soils are inherently variable, not only between sites, but both vertically within the soil column and spatially across short distances, a protocol study (Fournier et al., 1994) was conducted that compared three soil types: a till-derived Ferro-Humic Podzol (Haplorthod) at Turkey Lakes Watershed, Algoma District, Ontario; an alluvium-derived Humo-Ferric Podzol (Haplorthod), also from northern Ontario; and a Dystric Brunisol (Dystrochrept) from the southernmost part of Canada, near Lake Erie. Each profile was sampled at various depths across a grid, then analyzed by the same procedures used in the ARNEWS (below). The resulting data were used to estimate numbers of samples required for statistically valid comparisons (data for the Turkey Lakes soil are presented in Table 4–3). Sample numbers required varied from element-to-element and from horizon-to-horizon, and were generally lower in organic compared with mineral horizons and for macro- as opposed to microelements. Substantial caution, however, must always be exercised when interpreting results.

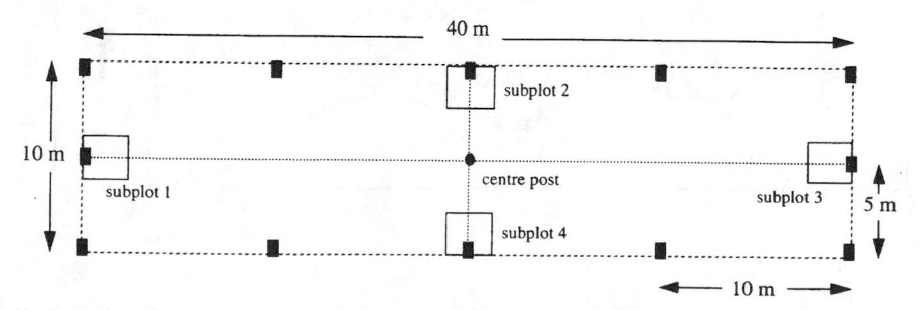

Fig. 4–1. ARNEWS plot configuration (from D'Eon et al., 1994).

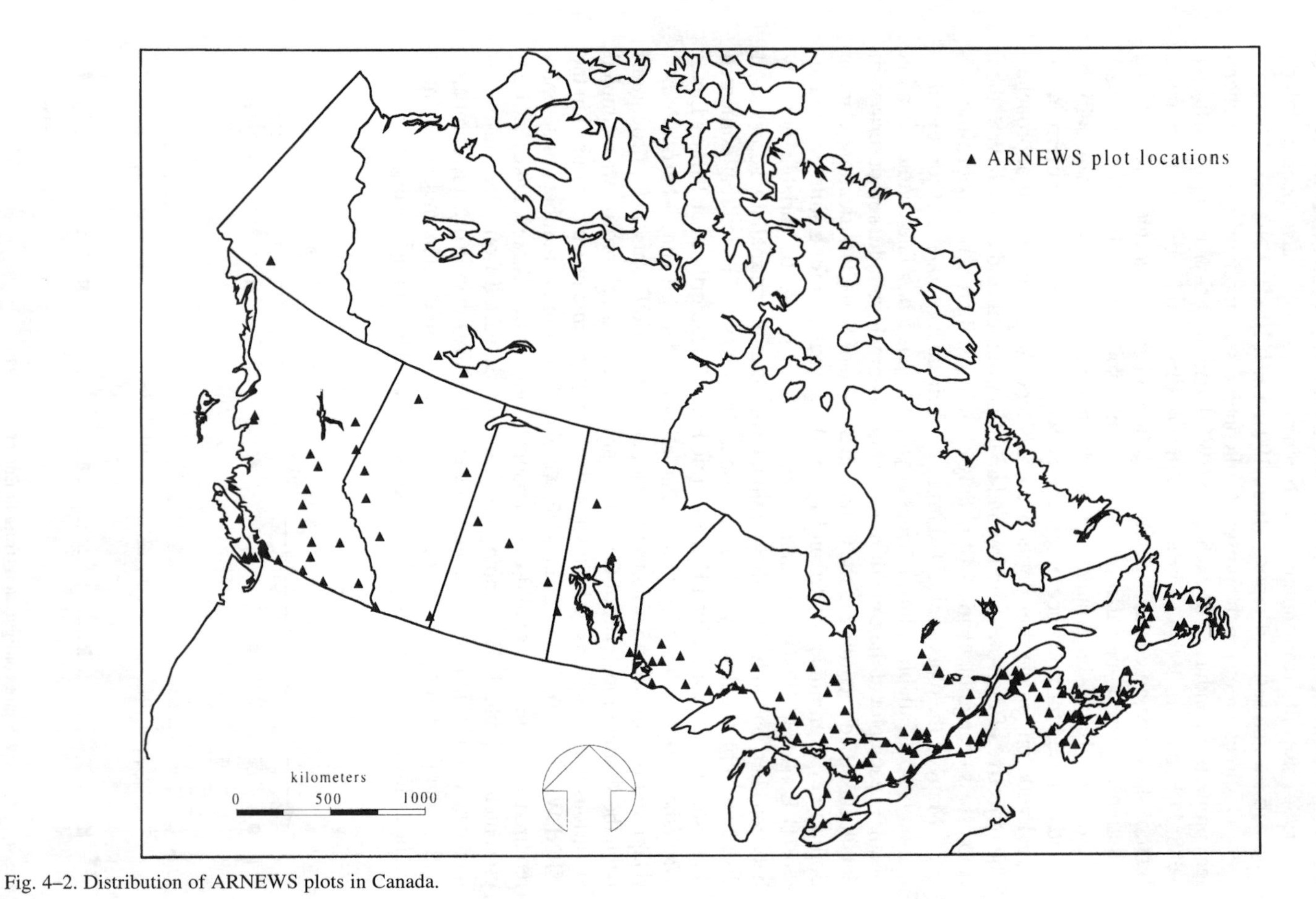

Fig. 4–2. Distribution of ARNEWS plots in Canada.

Table 4–2. ARNEWS parameter assessment schedule (reformatted from D'Eon et al., 1994).

Base year
- Plot characteristics
- Tree mapping

Several times per growing season
- Abiotic foliar symptoms
- Insect and disease conditions
- Seed production

Base year plus annual assessments
- Tree mortality in the stand
- Tree condition

Every second year
- Regeneration and saplings
- Ground vegetation

Base year plus 5-yr assessments
- Radial tree growth
- Vertical tree growth
- Crown structure and density
- Foliage sampling for analysis
- Soil sampling for analysis
- Softwood shoot growth

For long-term monitoring, a lesser number of soil horizons but at a large number of points is selected for periodic sampling and analysis. In practice, five stations are located around the periphery of each plot and, within 5 m of a center stake, 10 subsamples of F-(Förmultningsskiktet), and five each of the most abundant A-(eluviated), and B-(illuviated) horizons are collected and composited into F, A, and B station samples for analysis. Soil samples are air dried, screened to pass a 2-mm sieve, and analyzed using standard tests (Kalra & Maynard, 1991). These include pH (both in H_2O and $CaCl_2$), organic matter (loss-on-ignition; wet

Table 4–3. Sample numbers required to estimate mean results of chemical analyses ±10% with 95% confidence for a Ferro-Humic Podzol (Haplorthod) soil, Turkey Lakes Watershed, Ontario (selected from Fournier et al., 1994).

Parameter	Units	L	F	H	Ahe	Bhf
Total N	mg g^{-1}	4	3	28	93	31
Available P	μg g^{-1}	79	24	107	243	26
Exchangeable K	cmol kg^{-1}	14	20	31	54	77
Exchangeable Ca	cmol kg^{-1}	13	23	67	153	276
Exchangeable Mg	cmol kg^{-1}	12	15	45	72	148
Exchangeable SO_4	μg g^{-1}	38	36	20	44	20
Exchangeable Fe	cmol kg^{-1}	490	934	467	191	215
Exchangeable Mn	cmol kg^{-1}	40	64	243	330	673
Exchangeable B	μg g^{-1}	56	102	92	79	385
Exchangeable Zn	cmol kg^{-1}	71	24	99	108	531
Exchangeable Cu	cmol kg^{-1}	19	11	28	88	90
Exchangeable Mo	μg g^{-1}	78	85	69	65	51
Exchangeable Na	cmol kg^{-1}	21	26	23	19	21
Exchangeable Al	cmol kg^{-1}	380	156	238	78	68
Cation-exchange capacity	cmol kg^{-1}	19	19	73	33	25
Orgainc matter	%	9	7	71	112	31
Loss on ignition	%	3	5	51	130	52
pH($CaCl_2$)		1	1	2	1	1
pH(H_20)		1	1	2	1	1

oxidation optional), total N, total and available (Bray and Kurtz no. 1) P, total and exchangeable (unbuffered 1.0 *M* NH_4Cl) K, Ca, Mg, S, as well as a number of other elements.

AN EXAMPLE FROM THE ARNEWS SOIL RESAMPLING

An initial sampling of soils was conducted in 1985 (as late as 1987 in a few cases) and the first resampling (of the 103 plots established in 1985) was conducted in 1990 (a second round of resampling was carried out in 1995, but results are not yet available). The principal comparison of the ARNEWS soil monitoring program is between samples of soil from the same plot over time. To allow some insight into possible effects related to acid deposition, however, mean annual wet SO_4^{2-} and wet NO_3^- deposition values during the same 5-yr period (including the years 1985 to 1989) were assembled, mainly from Environment Canada sources, and mean depositions of SO_4^{2-} and NO_3^- for each plot location were determined by interpolation. Statistical analysis was made by, first, categorizing each plot into one of five deposition zones of SO_4^{2-} and of NO_3^- (Table 4–4), then analyzing soil concentration changes between 1985 and 1990 by analysis of variance and by Duncan's New Multiple Range Test for the detection of differences ($P = 0.05$) among means (Milliken & Johnson, 1984). Results for pH, exchangeable K^+, Ca^{2+}, and Mg^{2+} have been described (Morrison et al., 1996). To illustrate, however, results for Ca^{2+} and Mg^{2+}, two of the principal cations of the forest floor (F-horizon) and topmost mineral layer (A-horizon) are presented again. In addition to within-site variability, a further important reason for caution in interpreting these particular data, however, is potential confounding related to the overlay of deposition pattern (of SO_4^{2-} or NO_3^-) on either soil or forest distribution. The situation with regard to forest distribution is especially problematic in that the overwhelming number of Zone 1 to 3 plots (of both the SO_4^{2-} and NO_3^- series) support coniferous forest, whereas hardwood forest predominates in Zones 4 and 5. Also, Zones 4 and 5 in eastern Canada tend to be contiguous with more temperate climates and different bedrock geologies. Soil profile types, on the other hand, were fairly evenly distributed amongst the zones. Nevertheless, there appeared to be some correlation between soil properties and acid deposition.

F-horizon exchangeable Ca^{2+} concentrations (Fig. 4–3) were, on average, unchanged over the period 1985 to 1990 in less-impacted zones of both SO_4^{2-} and

Table 4–4. Deposition zones of SO_4^{2-} and NO_3^- deposition, 1985 to 1989 inclusive.

SO_4^{2-} Zone	SO_4^{2-} Deposition	NO_3^- Zone	NO_3^- Deposition
	kg ha^{-1} yr^{-1}		kg ha^{-1} yr^{-1}
1	0–9.9	1	0–4.9
2	10.0–14.9	2	5.0–9.9
3	15.0–19.9	3	10.0–14.9
4	20.0–24.9	4	15.0–19.9
5	25.0+	5	20.0+

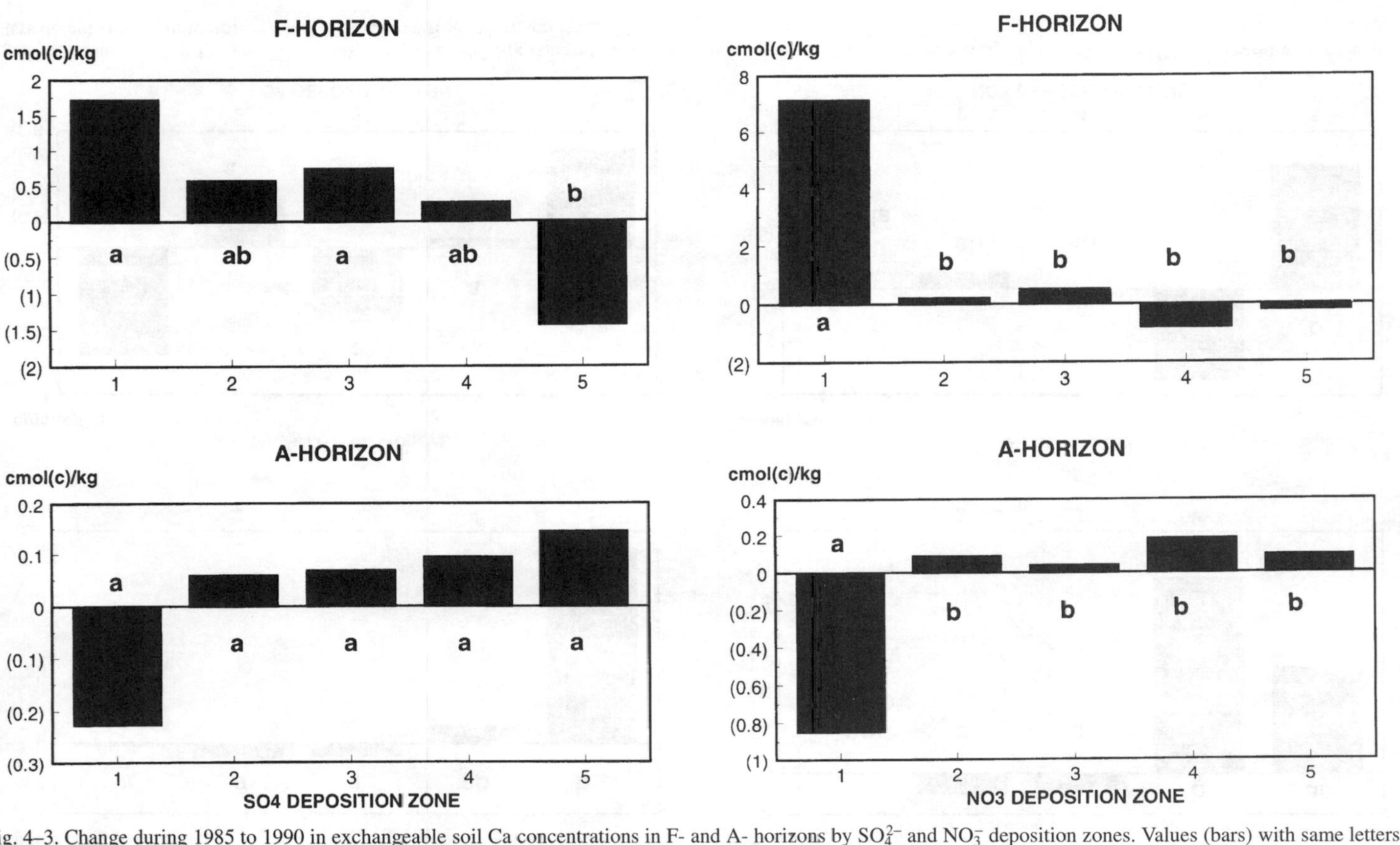

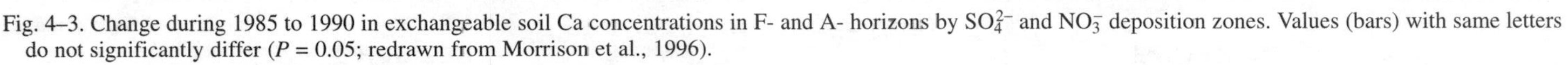

Fig. 4–3. Change during 1985 to 1990 in exchangeable soil Ca concentrations in F- and A- horizons by SO_4^{2-} and NO_3^- deposition zones. Values (bars) with same letters do not significantly differ ($P = 0.05$; redrawn from Morrison et al., 1996).

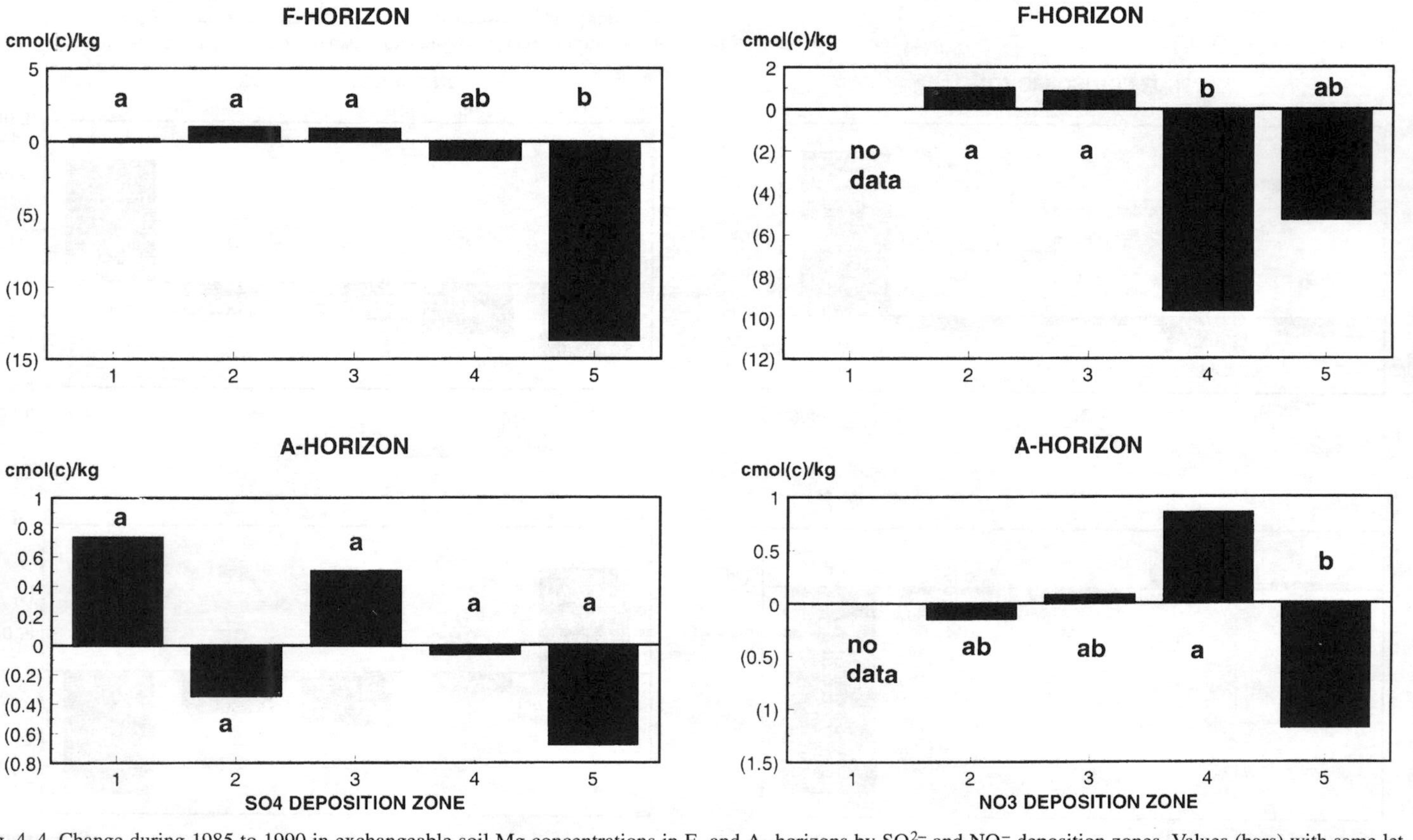

Fig. 4–4. Change during 1985 to 1990 in exchangeable soil Mg concentrations in F- and A- horizons by SO_4^{2-} and NO_3^- deposition zones. Values (bars) with same letters do not significantly differ ($P = 0.05$; redrawn from Morrison et al., 1996).

NO_3^- deposition, and lower in more-impacted zones. Regarding A-horizon exchangeable Ca^{2+}, similar though less-consistent trends were evident.

A similar situation appeared to exist with respect to 5-yr exchangeable Mg^{2+} change (Fig. 4–4), with higher forest floor Mg^{2+} concentrations in less-impacted zone (particularly Zone 1) soils in 1990 as opposed to 1985, vs. unchanged or lower concentrations in zones of higher deposition of both SO_4^{2+} and NO_3^-. The reverse, however, occurs in the topmost mineral horizon.

DISCUSSION

The overall goal of the C&I process is to gauge changes in the status of forests and related conditions over time (Canadian Forest Service, 1995) to assist in shaping sustainable management policy. A significant number of the Montreal Process indicators do concern current policy. Others, while mainly compilations of statistics, have obvious policy implications (e.g., extent of area in protected area categories, or annual removal of wood products compared with sustainable volume). Others, particularly those of Criteria 3 and 4, concern disturbance, with the greater number presupposing human disturbance. Included here are Indicators 3b and 4h that concern area and percentage of forest land subject to pollution.

In general, the purpose of monitoring is to establish baselines, then to determine departures from those baselines over time. The soils monitoring initiative of the ARNEWS and the soil pollution-related indicators of the C&I have at least this in common. Beyond this, there is some divergence. ARNEWS soil monitoring activities focus on *effects*, expressed as *change in soil properties* over time and, from this, possibly insight into *processes* at the stand level. From the above data, for example, it is tempting to interpret changes in concentrations of Ca^{2+} or Mg^{2+} as faint signals that soil bases are being leached from the forest floor and upper mineral soil under the influence of sulphuric or nitric acid rain. This would at least be consistent with the theory that H^+-ions displace base cations from exchange sites in the soil, and that SO_4^{2-} or NO_3^- ions, act as leaching counter-ions to transport them deeper into or even out of the soil (Johnson & Cole, 1977; Cronan et al., 1978; Cronan, 1980; Molliter & Raynal, 1982; Foster, 1985; Foster et al., 1986). Sampling and analysis of soil during subsequent 5-yr periods will be required, however, to determine if the trends observed during 1985 to 1990 continue. Even with observed trends, however, many factors may have contributed to the result and alternative explanations should not be ruled out. The Montreal Process pollution-related indicators, on the other hand, are stated in terms of *area*, and represent a landscape level approach. The first (Indicator 3b) is concerned with specific air pollutants but requires only that *area and percentage of forest land subject to specific air pollutants (e.g., sulphates, nitrates, ozone) or ultraviolet B that may cause negative impacts on the forest ecosystem* be ascertained. The current wording does not necessarily require that *effects* be measured. The other soil pollution-related indicator, concerned with *accumulations of persistent toxic substances*, is stated in a marginally-more effects-oriented manner, but still fails to give an adequate picture, though for a somewhat different reason. In the Sudbury smelter smoke example described earlier, Ni^{2+} and

Cu^{2+} accumulations in the affected soils were presumably the main phytotoxins, yet other soil attributes, principally availabilities of base cations, mainly K^+ and Ca^{2+} also were negatively affected by the depositions. Thus, while Ni^{2+} or Cu^{2+} accumulations might have been captured under Indicator 4h, the collateral changes would not have and, in some cases, these may have been significant. With metals, chemical form is important. Many highly-damaged soils in Sudbury area have, for example, been reclaimed in part with lime applications, which affects the chemical availabilities of Ni^{2+} and Cu^{2+} but has no effect on total amounts of these elements per se. Thus, some indication of chemical form is necessary as well.

Some general lessons from the ARNEWS experience may also be of value in indicator establishing and assessing. One is obvious. A system must be do-able and there are significant realities to be contended with in adapting landscape-level analyses of molecular-level processes to Canadian conditions. First, Canada has a substantial forest area to assess. Some 417.6 million ha are dominated by forest vegetation, with some 234.5 million hectares of this considered to be commercially-operable forest land (Canadian Forest Service, 1996). To further place this in perspective, the Canadian forest extends in terms of latitude, for example, from north of the Arctic Circle to roughly the same latitude as northern California, USA, or the Mediterranean Sea in Europe; and from east to west, through roughly 90° of longitude or 7600 km. Second, the forest is not uniform. One commonly-used classification (Rowe, 1972) divides Canada into eight distinct forest regions. In the east: the Acadian Forest Region, with species such as red spruce, balsam fir [*Abies balsamea* (L.) Mill.], red maple (*Acer rubrum* L.), sugar maple, and yellow birch (*Betula alleghaniensis* Britton). Further west is the Great Lakes-St. Lawrence Forest Region, dominated by sugar maple, yellow birch, beech (*Fagus grandifolia* Ehrh.), white pine (*Pinus strobus* L.), and red pine (*P. resinosa* Ait.). North of the lower Great Lakes in the middle of North America is a small strip of Carolinian vegetation, with species such as chestnut [*Castanea dentata* (Marsh.) Borkh.], black walnut (*Juglans nigra* L.), tulip-tree (*Liriodendron tulipifera* L.), and cucumber-tree (*Magnolia acuminata* L.) whereas 400 km to the north begins a Boreal Forest, dominated by jack pine (*Pinus banksiana* Lamb.), black spruce (*Picea mariana* (Mill.) B.S.P.), trembling aspen (*Populus tremuloides* Michx.), and white birch, which ultimately grades into Tundra. In the Rocky Mountains, there are the sub-Alpine, Montane, and Columbia Forest Regions, and on the Pacific coast, the Coast Forest Region, with Douglas-fir [*Pseudotsuga menziesii* (Mirb.) Franco], western hemlock [*Tsuga heterophylla* (Raf.) Sarg.], Sitka spruce [*P. sitchensis* (Bong.) Carr.], and western red cedar (*Thuja plicata* Donn). There are, in fact, in excess of 150 species that attain tree size in Canada (a number greater than the total number of ARNEWS plots currently in existence). Third, natural disturbances, such as wildfire or insect infestation in Canada can occur on a large scale. Wildfire, for example, during the 10-yr period ending in 1995 burned, on average, 2.5 million ha annually with, in 1994, this figure being 6.3 million ha (Canadian Forest Service, 1996). In 1994 as well, damaging insects infested some 11.6 million ha (ibid.). Thus, even routine inventory can place a significant demand on resources. Fourth, with specific reference to soils per se, on-the-ground soils mapping is limited, being restricted largely to areas where agriculture is conducted and is often of very limited availability in areas where forests predominate. Finally, in comparison to the total land area, Canada has a relatively small population and, while a major producer and exporter of forest products, forest management in Canada, in most cases, is extensive; and the number of resource professionals available for indicator assessment is small. It is unlikely, therefore, that in Canada a C&I pollution monitoring initiative would be undertaken independent of existing systems, the most significant of which is the ARNEWS.

The current ARNEWS has a clearly long-range pollution focus, though some ARNEWS-type installations are being considered for deployment around specific source areas as well (L. Turchenek, 1997, personal communication). Under the current ARNEWS, a wide range of ecosystem attributes, including a significant selection of soil properties are already periodically assessed, but others could be added if necessary. For example, in soil (or, for that matter, in foliage), certain polluting metals such as Ni, Cu, Pb, or Cd, are not routinely analyzed for, but could be with little additional effort. Similarly, organic substances also could analyzed for if necessary. Some pitfalls are evident from the ARNEWS soils experience, however, and may be of value to any C&I initiative. Two have been alluded to already: short-range variations in soil necessitating large numbers of sampling points, and the widespread occurrence of air pollution at low levels. Further, while field sampling and laboratory preparation and analysis of soils are conducted according to common protocols (e.g., Magasi, 1988), departures can be expected to occur. With large numbers of individuals in the field, operator-to-operator variation is difficult to avoid and, because some field operations (e.g., foliage sampling) should be conducted within relatively-narrow time constraints, reducing the number of operators is not always feasible. Attempts must be made, then, to minimize operator-to-operator differences through staff training and other quality control–quality assurance procedures. Likewise, while similar methodologies may be employed when analyzing samples for chemical constituents, both between-laboratory and within-laboratory baseline drift also can contribute to overall methodological error. Arguments are sometimes advanced in favor of methodological standardization; however, the counter argument is that methods standardized too rigidly also can become methods frozen in time. This is especially significant in forest monitoring because of the long growth periods of forest crops.

Reference methods, reference materials, sample archiving, and procedures run in parallel are probably of greater value. Even then, bias with what frequently may be perceived of as *old data* may be hard to overcome. Another area involves analytical methodology. Within forest soil science, progress has been slow in selecting or developing methodologies for the determination of soil chemical properties that have meaning for tree growth. Extraction methods developed and tested for agricultural soils and crops have, of necessity, been widely used for forest soils and crops, often with little testing. Thus, there is a virtually wide-open field for the development of soils monitoring techniques that have meaning in terms of tree response, yet at the same time be capable of widespread application.

CONCLUSIONS

The C&I invites landscape-level assessments. As currently stated, however, the two pollution-related indicators may not in themselves produce particularly meaningful information. For example, Indicator 3b, specifying *area and percentage of forest land subject to specific air pollutants* . . . could, if its wording were taken at face value, have its reporting requirement satisfied simply by overlaying deposition isopleths on a forest map of Canada and counting grid-squares. A rewording incorporating *effects* would greatly increase its usefulness. Likewise, Indicator 4h should be modified to emphasize availability or chemical form in addition to accumulation. A better wording for Montreal Indicator 3b, therefore, might be *area and percentage of forest land exhibiting significant changes in soil chemical, biological, or physical properties attributable to specific air pollutants (e.g., sulphates, nitrates, ozone) or ultraviolet B that may cause negative impacts in forest ecosystem* and a better rendering of Indicator 4h, might be *area and percentage of forest land experiencing accumulations of persistent toxic substances in the soil in forms likely to cause negative impacts in forest ecosystems.* This would enable an interface of C&I with its area and percentage of emphasis with the ARNEWS properties emphasis and results from other more-intensive process-related studies.

The Canadian forest situation may have little in common with some countries, but have significant parallels in others. The chief relevant features of the Canadian situation are a very large area to assess and a very small number of resource professionals to do the assessing. Under such circumstances, it would, for the pollution-related indicators, be preferable to build on existing data bases such as the ARNEWS in Canada or similar networks in other countries. Required, then, are acceptable methods of scaling plot-type data up to the landscape level.

Expansion of the ARNEWS or ARNEWS-like networks with, perhaps, a monitoring of additional parameters could yield insights at the properties–process level in relation to criteria other than the two pollution-related criteria emphasized in this chapter. Process-level work could yield insights into the fundamental ecological processes aspects of Montreal Indicator 3c or Indicators a and b of Criterion 5 (Maintenance of Forest Contribution to Global Carbon Cycles).

ACKNOWLEDGMENTS

The author is indebted to R.E. Fournier, Canadian Forest Service, Sault Ste. Marie, Ontario, for assistance in compiling ARNEWS soil analytical data, and to Atmospheric Environment Service of Environment Canada, Downsview, Ontario, for access to and assistance in compiling sulphate and nitrate deposition data.

REFERENCES

Addison, P.A. 1989. Monitoring the health of a forest: a Canadian approach. Environ. Monitor. Assess. 12:39–48.

Addison, P.A., K.A. Kennedy, and D.G. Maynard, 1984. Effects of sour gas processing on a forest ecosystem in west-central Alberta. Inf. Rep. NOR-X-265. Can For. Serv., Edmonton, AB.

Amiro, B.D., and G.M. Courtin, 1981. Patterns of vegetation in the vicinity of an industrially disturbed ecosystem, Sudbury, Ontario. Can. J. Bot. 59:1623–1639.

Baker, J. 1977. Nutrient levels in rainfall, lodgepole foliage, and soils surrounding two sulfur gas extraction plants in Strachan, Alberta. Inf. Rep. NOR-X-194, Can For. Serv., Edmonton, AB.

Bertrand. A., G. Robitaille, P. Nadeau, and R. Boutin, 1994. Effects of soil freezing and drought stress on abscisic acid content of sugar maple sap and leaves. Tree Physiol. 14:413–425.

Blauel, R.A., and D. Hocking, 1974. Air pollution and forest decline near a nickel smelter. The Thompson, Manitoba Smoke Easement Survey, 1972–1974. Inf. Rep. NOR-X-115, Can. For. Serv., Edmonton, AB.

Canada Soil Survey Committee. 1978. The Canadian system of soil classification. Publ. 1646. Can. Dep. Agric., Ottawa, ON.

Canadian Council of Forest Ministers. 1995. Sustainable forest management. A Canadian approach to criteria and indicators. Can. Forest Serv., Ottawa, ON.

Canadian Forest Service. 1995. Criteria and indicators for the conservation and sustainable management of temperate and boreal Forests. The Montreal Process. Can. Forest Serv., Hull, QC.

Canadian Forest Service. 1996. The state of Canada's Forests 1995–1996. Can. For. Serv., Ottawa, ON.

Canadian Standards Association. 1996. A sustainable forest management system: Guidance document. Environmental management systems. Can. Standards. Assoc. Etobicoke, ON.

Canadian Standards Association. 1996. Sustainable Forest Management System: Specifications Document. Environmental management systems. Can. Standards. Assoc., Etobicoke, ON.

Carrier, L. 1986. Decline in Quebec's forests. Assessment of the situation. Serv. de la recherche appliquée. Min. de l'Energie et des Ressources. Québec, QC.

Carrier, L., and G. Gagnon, 1985. Maple dieback in Quebec. Serv. de la recherche appliquée. Min. de l'Energie et des Ressources. Québec, QC.

European Commission, Diretorate-General for Agriculture. 1989. European community forest health report. EC, Brussels.

European Commission, Diretorate-General for Agriculture. 1990. European community forest health report. EC, Brussels.

European Commission, Diretorate-General for Agriculture. 1991a. European community forest health report. EC, Brussels.

European Commission, Diretorate-General for Agriculture. 1991b. European community forest health report. EC, Brussels.

Cronan, C.S. 1980. Solution chemistry of a New Hampshire subalpine ecosystem. Oikos 34:272–281.

Cronan, C.S., W.A. Reiners, R.C. Reynolds, and G.E. Lang, 1978. Forest floor leaching: Contributions from mineral, organic and carbonic acids in New Hampshire subalpine forests. Science (Washington, DC) 200 (4339):309–311.

D'Eon, S.P., L.P. Magasi, D. Lachance, and P. DesRochers, 1994. ARNEWS. Canada's national forest health monitoring plot network manual on plot establishment and monitoring. Revised. Inf. Rep. PI-X-119, Can. For. Serv., Chalk River, ON.

D'Eon, S.P., and J.M. Power, 1989. The acid rain national early warning system (ARNEWS) plot network. Inf. Rep. PI-X-91, Can. For. Serv., Chalk River, ON.

Dreisinger, B.R. 1970. SO_2 levels and vegetation injury in the Sudbury area during the 1969 season. Ontario Dep. Energy Resour. Manage., Toronto, ON.

Foster, N.W. 1985. Acid precipitation and soil solution chemistry within a maple-birch forest in Canada. For. Ecol. Manage. 12:215–231.

Foster, N.W., I.K. Morrison, and J.A. Nicolson, 1986. Acid deposition and ion leaching from a podzolic soil under hardwood forest. Water Air Soil Pollut. 31:879–890.

Fournier, R.E., I.K. Morrison, and A.A. Hopkin, 1994. Short range variability of soil chemistry in three acid soils in Ontario, Canada. Commun. Soil Sci. Plant Anal. 25:3069–3082.

Freedman, B., and T.C. Hutchinson. 1980. Pollutant inputs from the atmosphere and accumulations in soils and vegetation near a nickel-copper smelter at Sudbury, Ontario, Canada. Can. J. Bot. 58:108–132.

Glass, N.R., G.E. Glass, and P.J. Rennie, 1979. Effects of acid precipitation. Environ. Sci. Technol. 13:1350–1355.

Glass, N.R., G.E. Glass, and P.J. Rennie, 1980. Effects of acid precipitation in North America. Environ. Int. 4:443–452.

Gordon, A.G., and E. Gorham, 1963. Ecological aspects of air pollution from an iron-sintering plant at Wawa, Ontario. Can. J. Bot. 41:1063–1078.

Gorham, E., and A.G. Gordon, 1960. Some effects of smelter pollution northeast of Falconbridge, Ontario. Can. J. Bot. 38:307–312.

Hall, J.P. 1991. ARNEWS annual report 1990. Inf. Rep. ST-X-1, Can. Forest. Serv., Ottawa, ON.

Hall, J.P. 1993. ARNEWS annual report 1992. Inf. Rep. ST-X-7, Can. Forest Serv., Ottawa, ON.

Hall, J.P., and B.A. Pendrel, 1992. ARNEWS annual report 1991. Inf. Rep. ST-X-5, Can. Forest. Serv. Ottawa, ON.

Hendershot, W.H., and A.R.C. Jones, 1989. Maple decline in Quebec: A discussion of possible causes and the use of fertilizers to limit damage. For. Chron. 65:280–287.

Hogan, G.D., and D. Wotton, 1984. Pollutant distribution and effects on forests adjacent to smelters. J. Environ. Qual. 13:377–381.

Hutchinson, T.C., and L.M. Whitby, 1974. Heavy-metal pollution in the Sudbury mining and smelting region of Canada: I. Soil and vegetation contamination by nickel, copper, and other metals. Environ. Conserv. 1:123–132.

Johnson, D.W., and D.W. Cole. 1977. Anion mobility in soils: Relevance to nutrient transport from terrestrial to aquatic systems. USEPA600/3-77-068. U.S. Environ. Protection Agency, Corvallis, OR.

Kalra, Y.P., and D.G. Maynard, 1991. Methods manual for forest soil and plant analysis. Inf. Rep. NOR-X-319, Forest. Can., Edmonton, AB.

Lachance, D., A. Hopkin, B. Pendrel, and J.P. Hall, 1995. Health of sugar maple in Canada. Results from the North American Maple Project, 1988–1993. Inf. Rep. ST-X-10, Can. For. Serv., Ottawa, ON.

Leblanc, F., G. Robitaille, and R.N. Rao, 1974. Biological response of lichens and bryophytes to environmental pollution in the Murdochville copper mine area, Quebec. J. Hattori Bot. Lab. 38:405–433.

Linzon, S.N. 1958. The influence of smelter fumes on the growth of white pine in the Sudbury region. Ontario Dep. Lands For./Ontario Dep. Mines Publ., Toronto, ON.

Lozano, F.C., and I.K. Morrison, 1981. Disruption of hardwood nutrition by sulphur dioxide, nickel, and copper air pollution near Sudbury, Canada. J. Environ. Qual. 10:198–204.

Magasi, L.P. 1988. Acid rain national early warning system manual on plot establishment and monitoring. Inf. Rep. DPC-X-25, Can. Forest. Serv., Ottawa, ON.

Magasi, L.P. 1989. White birch deterioration in the Bay of Fundy region, New Brunswick 1979–1988. Inf. Rep., Forest. Can., Fredericton, NB.

McLaughlin, D.L., S.N. Linzon, D.E. Dimma, and W.D. McIlveen, 1985. Sugar maple decline in Ontario. Ontario Rep. ARB-144-85-Phyto., Ontario Min. Environ., Toronto, ON.

Milliken, G.A., and D.E. Johnson, 1984. Analysis of messy data: I: Designed experiments. Lifetime Learning Publ., Belmont, CA.

Molliter, A.V., and D.J. Raynal, 1982. Acid precipitation and ionic movements in Adirondack forest soils. Soil Sci. Soc. Am. J. 45:137–141.

Morrison, I.K. 1984. Acid rain: A review of acid deposition effects on forest ecosystems. For. Abstr. 44:483–506.

Morrison, I.K., R.E. Fournier, and A.A. Hopkin, 1996. Response of forest soil to acid deposition: Results of a 5-year resampling study in Canada. p. 187–197. *In* R. Cox et al.(ed.) Air pollution and multiple stresses. IUFRO, 16th Annual Meeting for Specialists in Air Pollution. Effects on Forest Ecosyst, Fredericton, NB. 7–9 Sept. 1994. Can. For. Serv., Atlantic Cen., Fredericton, NB.

National Research Council of Canada. Associate Committee on Trail Smelter Smoke. 1939. Effect of sulphur dioxide on vegetation. Natl. Res. Counc. Can., Ottawa, Ontario Publ. Na815, NRC, Ottawa.

Robitaille, G, .F. Leblanc, and D.N. Rao, 1977. Acid rain: A factor contributing to the paucity of epiphytic cryptograms in the vicinity of a copper smelter. Rev. Bryol. Lichenol. 43:53–66.

Rowe, J.S. 1972. Forest regions of Canada. Publ. 1300, Dep. Environ., Can. For. Serv., Ottawa, ON.

Sidhu, S.S., and B.A. Roberts. 1976. Progression of fluoride damage to vegetation from 1973 to 1975 in the vicinity of a phosphorus plant. Bi-Mon. Res. Notes, Can. For. Serv. 32:29–31.

Sidhu, S.S., and R.J. Staniforth, 1986. Effects of atmospheric fluorides on foliage and cone and seed production in balsam fir, black spruce and larch. Can. J. Bot. 64:923–931.

Tamm, C.O., and E.B. Cowling, 1977. Acidic precipitation and forest vegetation. Water Air Soil Pollut. 7:503–511.

Tomlinson, G.H. 1983. Die-back of red spruce, acid deposition, and changes in soil nutrient status: A review. p. 331–342. *In* B. Ulrich and J. Pankrath (ed.) Effects of accumulation of air pollutants in forest ecosystems. D. Reidel Publ. Co., Dordrecht, the Netherlands.

Ulrich, B., R. Mayer, and P. Khanna, 1980. Chemical changes due to acid precipitation in a loess-derived soil in central Europe. Soil Sci. 130:193–199.

United Nations-Economic Commission for Europe, and European Commission. 1992. Forest condition in Europe. Report on the 1991 survey. UN-ECE/EC, Brussels, Geneva.

United Nations-Economic Commission for Europe, and European COMMISSION. 1993. Forest condition in Europe. Report on the 1992 survey. UN-ECE/EC, Brussels, Geneva.

United Nations-Economic Commission for Europe, and European COMMISSION. 1994. Forest condition in Europe. Report on the 1993 survey. UN-ECE/EC, Brussels, Geneva.

United Nations-Economic Commission for Europe, and European COMMISSION. 1995. Forest condition in Europe. Report on the 1994 survey. UN-ECE/EC, Brussels, Geneva.

United Nations-Economic Commission for Europe, and European COMMISSION. 1996. Forest condition in Europe. Report on the 1995 survey. UN-ECE/EC, Brussels, Geneva.

Vogelmann, H.W. 1982. Catastrophe on Camels Hump. Nat. Hist. 91:8–14.

Whitby, L.M., and T.C. Hutchinson, 1974. Heavy-metal pollution in the Sudbury mining and smelting region of Canada: II. Soil toxicity tests. Environ. Conserv. 1:191–200.

White, L.T., and S.N. Linzon, 1950. Preliminary report on the condition of white pine in the Sudbury sulphur-fume area. Dominion For. Path. Lab., Toronto, ON.

5 Criteria and Indicators of Acceptable Atmospheric Deposition of Sulfur and Nitrogen on Forests in Western Europe

Nico Van Breemen

Department of Soil Science and Geology
Wageningen Agricultural University
Wageningen, the Netherlands

Wim De Vries

DLO-Winand Staring Centre
Wageningen, the Netherlands

Pieter H.B. De Visser

AB-DLO
Wageningen, the Netherlands

As stated in the Montreal Process (Anonymous, 1995), the "development of internationally agreed criteria and indicators for the conservation and sustainable management of forests" is an important step to guide policy-makers in the formulation of policies aimed at supporting sustainability of forests. One topic where the development of such criteria has received a lot of interest during the last decade is that of atmospheric deposition of S and N. In this context, the concept of critical loads has been introduced (Nilsson, 1986). A critical acid load has been defined as "the maximum deposition of acidifying compounds that will not cause chemical changes leading to long-term harmful effects on ecosystem structure and function" (Nilsson & Grennfelt, 1988). Critical acid loads for forests have thus been derived using relatively simple steady-state models (e.g., De Vries, 1993; Hettelingh et al., 1991; De Vries et al., 1994). Critical N loads for forests have been derived by using both empirical data and models during the last decade (e.g., Grennfelt & Thörnelöf, 1992; Bobbink et al., 1995; Hornung et al., 1995).

Elevated atmospheric deposition of S (1–5 kmol ha^{-1} yr^{-1}) and N (1.2–5 kmol ha^{-1} yr^{-1}) has strongly acidified most noncalcareous forest soils across western and central Europe. The same is true for forest streams and lakes in areas

The Contribution of Soil Science to the Development of and Implementation of Criteria and Indicators of Sustainable Forest Management. SSSA Special Publication no. 53.

with shallow, noncalcareous soil. Acidification has affected forest ecosystems through (i) decline of mycorrhizal fungi (Arnolds, 1991), (ii) shifts towards nitrophilic understory vegetation (Tyler, 1987; Van Dobben et al., 1999; Thimonier et al., 1994), (iii) yellowing and decreased retention of coniferous trees' needles (Hauhs & Wright, 1986; Schulze, 1989; Schütt & Cowling, 1983), (iv) reduced breeding success of birds due to declined supply of $CaCO_3$ from snails for bird egg-shell formation (Graveland et al., 1994), and (v) decline of aquatic life in forest streams (Arts, 1990; Henriksen & Hesthagen, 1993). Trees have declined locally, presumably by direct negative effects of elevated SO_2 on foliage, and from increased susceptibility to drought, frost, pests, and diseases, resulting from reduced root growth in acidified soil and from excessive N supply; however, across western Europe, growing stocks and tree growth have generally increased, presumably in part due to increased availability of N (Kauppi et al., 1992).

The research input into the acidification problem in Europe has been staggering. In the Netherlands alone total expenditure in acidification research in the past 15 yr was about US \$ 75×10^6. For the whole of Europe this figure must be in the order of US \$ 1×10^9. Probably more than one-half of the effort was directed to acidification of (semi-)natural terrestrial ecosystems, largely forests.

The aim of this chapter is to compare the Criteria and Indicator approach of the Montreal Process (Anonymous, 1995) with the European approach to derive environmental policy options by identifying critical loads for atmospheric deposition. This exercise may help to improve criteria related to the acidification problem, and hopefully provides some insight into the problems likely to be encountered when designing and applying criteria and indicators for other aspects of the conservation and sustainable management of such complex ecosystems as forests.

After an overview of the main changes in soil chemistry resulting from excessive soil acidification, the Montreal criteria and indicators most relevant to acidification will be discussed one by one in the light of research in Europe during the past 15 yr. While emphasis will be on the situation in the Netherlands, we will try to maintain a more general European perspective by including results from elsewhere.

SOIL ACIDIFICATION FROM ATMOSPHERIC DEPOSITION

Because the acidifying effects of N deposition cannot always be separated clearly from eutrophication by N, soil acidification here includes inadvertent fertilization. Acidified surface soils have a $pH(H_2O)$ of 3 to 4, and inorganic dissolved aluminum (Al^{3+}) as the major exchangeable and dissolved cation (De Vries et al., 1995c). Decreased supply of base cations and increased supply of Al is borne out by results from repeated sampling and analysis in the 1980s of southern Swedish forest soils that were studied one to six decades earlier (Falkengren-Grerup et al., 1987; Tamm & Hallbäcken, 1988). Similarly dissolved N, especially NO_3^- and Al^{3+}, markedly increased during periods of 6 to 20 yr of soil solution monitoring (Stein & van Breemen, 1993; Mulder & Stein, 1994). The causal

Table 5–1. Montreal Criteria relevant for acidification. Numbers in brackets refer to those of the Montreal Criteria (Ramakrishna & Davidson, 1998, this volume).

Compartment	Criteria†	Indicators
Tree layer	Maintenance of productive capacity (2) Maintenance of health and vitality (3)	Annual wood removal (d) Natural stress factors (a) Specific air pollutants (b) Biological process (c)
Understory	Maintenance of biological diversity (1)	Ecosystem and species fauna diversity (a, b)
	Maintenance of biological diversity (1)	Genetic diversity (c)
Soil, groundwater surface water	Conservation and maintenance of soil and water resources (4)	Soil chemical properties (b) Biological diversity of water bodies (f) Chemistry of water bodies (g)

† Indicators relevant for acidification are encountered under the Criterias 1 to 4.

relationship between elevated atmospheric deposition and soil acidification is illustrated by the fact that dissolved Al is balanced mainly by sulfate and NO_3^-. The concentrations of sulfate in the soil solution are directly related to the atmospheric input of S: forest soils are almost invariably SO_4-saturated (atmospheric input = drainage output) throughout Northwestern Europe. Where N deposition is very high, as in the Netherlands and adjacent areas in Germany and Belgium, soils are locally N-saturated according to the same criterion (atmospheric input = drainage output).

CRITERIA LIKELY TO BE RELEVANT FOR ACIDIFICATION

An overview of the criteria listed in the Montreal Process relevant with respect to acidification effects on forest ecosystem compartments considered is given in Table 5–1.

Criterion 1. Conservation of Biological Diversity

Indicator a, b: Ecosystem and Species Diversity

The indicator units considered under Ecosystem diversity are forest types, which are categories of forest defined mainly by their vegetation composition. These have clearly been affected by acidification in Europe, at least through shifts in nontree species composition.

In the understory vegetation of the Dutch forests, N deposition levels above 20 kg N ha^{-1} yr^{-1} have caused the disappearance of species of nutrient-poor habitats, e.g., the vascular plants Coral root bittercress [*Cardamine bulbifera* (L.) Crantz], Lungwort [*Pulmonaria officinalis* (L.) All.], common Solomon's seal (*Polygonatum multiflorum* L.), cross-leaved heath (*Erica tetralix* L.), and heather (*Calluna vulgaris* L.) and reindeer mosses (*Cladonia* sp., *Cladina* sp.) and the moss *Hylocomium splendens*. They were replaced by nitrophilous species, mainly wavy hair grass (*Deschampsia flexuosa* L.) and raspberry (*Rubus idaeus* L.) species (Van Dobben, 1993). As a result, the traditional poor pine forest types (Cladonia-Pinetum, Empetro-Pinetum) disappeared. Ecosystem diversity has diminished further by the change of heath land, as forest understory and as independent shrub ecosystem, towards grass-dominated ecosystems (Van der Eerden et al., 1995, in press). In stands of *Quercus robur* in Sweden, nitrophilous species like *Urtica dioica*, *Epilobium augustifolium*, and *Rubus idaeus* were more common at relatively high (12–15 kg N ha^{-1} yr^{-1}) than at low (6–8 kg N ha^{-1} yr^{-1}) N input (Tyler, 1987). In the last 35 yr in southern Sweden a number of plant species in oak and beech stands disappeared where the pH dropped below a certain value; however, in Sweden the total number of forest understory species has actually increased, due to the appearance of new N indicators species (Falkengren-Grerup & Eriksson, 1990). In northeastern France a similar increase in nitrophilous ground flora on calcareous soils was observed between two surveys (in 1972 and 1991; Thimonier et al., 1994).

In Germany and the Netherlands ectomycorrhizal fungi in mature forests decreased by almost 50% in the last 40 yr (Arnolds, 1991). The cause of decreased ectomycorrhizal fungi diversity is most probably increased N availability, not increased soil acidification. Fruit-body production of ectomycorrhiza is more strongly affected by elevated N than mycorrhizal infection, which may still be normal at inputs as high as 60 kg N ha^{-1} yr^{-1} (Branderud, 1995; Termorshuizen, 1990).

Empirical data indicate that to preserve the vegetation type prevalent before the advent of atmospheric eutrophication, N deposition should not exceed 15 to 20 kg ha^{-1} yr^{-1} , while deposition should remain below 7 to 20 kg ha^{-1} yr^{-1} to avoid the extinction of the most vulnerable species (Bobbink et al., 1995).

Indicator c: Genetic Diversity

Genetic diversity can possibly alter due to excessive soil acidification by extinction of acid-sensitive species or subspecies (genotypes). This follows from the well-known fact that acid tolerance can vary considerably within one plant species (Foy, 1988; Tan & Keltjens, 1990). In forests, only trees of an acid-tolerant genotype might survive years or decades of increased soil acidification. Birds too are affected by soil acidification. Breeding success of the Great tit (*Passer major* L.) was 40% due to egg shell failure where their source of Ca, snails, disappeared from increasingly acid, Ca-poor soils in extended forested areas (Graveland et al., 1994). The genetic diversity of these birds is at risk, due to disappearance of endemic populations and continuous replacement by immigrants from other ecosystems.

A related problem is the direct effect of elevated SO_2 on epiphytic lichens. Epiphytic lichens sensitive to SO_2 have disappeared over much of their former range in Europe (see e.g., Kuusinen et al., 1990), and while nitrophilous species have increased, e.g., in the Netherlands (Van Dobben, 1993) SO_2 sensitive lichens have partly recovered after atmospheric SO_2 concentrations decreased in the 1980s.

Criterion 2: Maintenance of Productive Capacity of Forest Ecosystems

Indicator d: The Annual Removal of Wood Products

In the early 1980s needle loss and yellowing in conifers was observed in 20 to 25% of central European forests, and local mortality of trees and stands was reported in areas with high emissions of S and/or N (Lammel, 1984; Schulze, 1989); however, Kauppi et al. (1992), estimated that only 8000 km^2, or 0.5% of the European forested area, was affected by severe decline. This by itself could hardly influence total forest resources and productivity. But in fact, between 1960 and 1990, growing stocks increased by 25% and forest growth by 30%, probably in part due to increased atmospheric N inputs (Kauppi et al., 1992). These authors concluded that the current trends of growing stocks is not likely to change in the coming decade or so, but that the favorable development of forest resources is at risk in the longer term by ongoing soil acidification. One of the indicators of productive capacity is the annual removal of wood products thought to be sustainable. While elevated N inputs may, at least temporarily, increase forest growth they will also increase the demand for base cation nutrients (Ca, Mg, K).

In the short-term, this increased demand may be fulfilled by base cation release from the exchange complex due to acidification; however, in strongly acidified soils, where the base saturation is nearly negligible, the supply of base cations for incremental growth is only by weathering and deposition. In a long-term perspective, sustainable forest growth is thus determined by the long-term supply of base cations through these processes. Base cation budgets made for Swedish forests indicate that the present uptake exceeds the input by weathering and deposition (Olsson et al., 1991). This implies that the present relatively fast growth of these forests is not sustainable from a long-term perspective.

Even in areas where the input of base cations is high, the high input of N may ultimately reduce forest growth due to effects on forest health (crown condition). For example, the availability of base cations may be reduced by increased levels of NH_4 and Al. This may cause an increased defoliation and discoloration (Criterion 3), which are often considered as signs of decreased vitality. A recent inventory of the needle composition of 150 forest stands in the Netherlands revealed that >60% of the stands suffered from a relative Mg deficiency due to high N contents (Hendriks et al., 1994). Based on a literature review, Roberts et al. (1989) also concluded that the decline in the vitality of spruce in central Europe is mainly due to foliar Mg deficiency.

Criterion 3: Maintenance of Forest Ecosystem Health and Vitality

Indicator a: Natural Stress Factors

The increased availability of N has increased the vulnerability of trees to drought, frost, pathogens and insect attacks (Van der Eerden et al., 1995, in press). The higher N content of leaf tissue prolongs the growing period and increases frost sensitivity by retarding the start of winter-hardening (Dueck et al., 1990). Corsican pine (*Pinus nigra poiretiana*) showed a larger number and size of infections with the fungus *Sphaeropsis sapinea* at wider N/K ratios in needle tissue, resulting from fertilization with ammonium sulphate (De Kam et al., 1991). Essentially all Corsican and Austrian pines (*Pinus nigra*) in heavily N-polluted areas in the Netherlands died from *Sphaeropsis*. Most forested areas in the Netherlands, Belgium, northeastern France, northwest Germany, Denmark, and southern Sweden have N inputs that are above their traditional biomass accumulation rate. This can be illustrated by the observed increases in biomass growth in the last decades (Kauppi et al., 1992). The greater aboveground biomass will increase the susceptibility of these forests to the traditional stress factors mentioned above, and also may increase the risk of wind-throw and drought (De Visser et al., 1996).

Indicator b: Specific Air Pollutants

The negative impacts of specific air pollutants on forest ecosystems relate to most aspects of Criteria 1 to 4. The influence of the pollutants N and S via soil on tree health is exerted by means of soil acidification and eutrophication. Concentrations of dissolved Al and soil pH now frequently exceed their no-effect level on tree roots, as demonstrated for seedlings in greenhouse studies. Effects observed in the field include shallow root systems with decreased branching (Marschner, 1991) and drastically reduced ectomycorrhizal infection (Arnolds, 1991). Supply of base cations has decreased by (i) loss of exchangeable cations due to soil acidification, (ii) antagonisms of cations with Al and NH_4 in soil, and (iii) diminished root growth and functioning; however, while soil acidification in the Netherlands has been quite strong, there is no evidence that this by itself has increased tree mortality, needle-shedding, or even growth (Hendriks et al., 1994).

Nitrogen is often the most important growth-determining nutrient for plants in temperate forest. Therefore increased N availability might indeed positively influence growth and health of plants. In soil, only very high concentrations of NH_4–N are directly harmful. Such negative effects include ion antagonism during uptake, and acidification due to nitrification, in addition to the effects of excessive N in foliage described under Indicator a.

In conclusion, tree health in most of the forests in northwestern Europe is influenced by N and S deposition via soil processes, but this has not necessarily decreased tree growth. Especially in areas with strong air pollution due to local emissions, tree growth may be affected by direct effects of e.g., SO_2, O_3, and NH_3 on tree canopies, but these are beyond the scope of this chapter.

Indicator c: Biological Processes

Acid deposition has clearly affected fundamental biological and ecological processes (e.g., soil nutrient cycling) and related ecological attributes (functionally important organisms such as mycorrhizal fungi) in European forest ecosystems; however, in the Netherlands, where acidification and increased Al supply have been more pronounced than in most other countries, tree vitality (as indicated by defoliation class) showed no correlation with dissolved Al in soils (Hendriks et al., 1994). This contrasts with the hypothesis, based mainly on laboratory and greenhouse evidence, that increased soluble Al in soils decreases tree vitality (Ulrich & Matzner, 1983; Sverdrup & Warfvinge, 1993).

Criterion 4: Conservation and Maintenance of Soil and Water Resources

Several of the indicators listed under this heading are relevant for acidification. They include changes in soil chemical properties (Indicator d), including organic matter pools, and changes in chemistry and biological diversity of water bodies in forest areas (Indicators f and g), and perhaps, in extreme cases, erosion.

Indicator d: Soil Chemical Properties

Changes in soil acidity status and nutrient supply have been dealt with under Criterion 3. Little if any research has been done on specific effects of acidification on soil organic matter. Increased acidity depresses decomposition of litter and humus, and increased forest productivity due to N deposition increases supply of litter (Aber et al., 1989), both of which tend to increase, not decrease, soil organic matter pools (Billet et al., 1990; Mallkönen et al., 1990). On the other hand, increased availability of N might stimulate litter decomposition, whereas a decreased availability of base cations due to soil acidification may ultimately decrease forest productivity. Definite predictions regarding trends in soil organic matter pools can thus not be given.

Indicator f: Biological Diversity of Waterbodies

Increased acidification of surface water and associated shifts in aquatic life, including decline of salmonoid fish, has been very extensively researched, and is relatively well understood. Large scale decline of salmonoid fish populations, particularly in southern Scandinavia and England, is directly caused by elevated dissolved Al due to acidification by atmospheric deposition (Hendriksen & Hesthagen, 1993; Havas & Rosseland, 1995).

Indicator g: Chemistry of Waterbodies

Streams and lakes in forested areas are particularly vulnerable to acidification. First, forests are usually confined to lower order watersheds, where water tends to be more acidic because it had little time for buffering by interaction with minerals. Furthermore adjacent soils tend to be relatively shallow and acidic (Driscoll et al., 1987), compared with downstream areas that are more likely to be used for agriculture. Finally, elevated dry deposition on forest canopies, particularly at higher elevations, usually leads to higher total acid loads in forested than in nonforested land (Hultberg, 1985).

Groundwater quality in forested areas has been widely affected by acidification as indicated by (i) decreased pH (Boumans & Beltman 1991), and (ii) elevated concentrations of Al (Lükewille & Van Breemen, 1992) and NO_3 (Boumans, 1994). These changes threaten actual and potential supplies of drinking water from aquifers under forested areas, that used to be renown for their good quality. Increased erosion can be expected in areas with widespread tree mortality, but we are not aware of reports of actual cases of such erosion.

CRITICAL LOADS OF ACID DEPOSITION IN USE IN EUROPE

Critical loads have been based on criteria derived from (i) empirical data that directly relate loads to effects and (ii) steady-state soil models that calculate critical loads from critical chemical values for ion concentrations or ratios in foliage, soil solution, and groundwater (De Vries, 1993).

Criteria to set critical loads in large parts of Europe, including England, Scandinavia, the Netherlands and Germany include (i) maximum soil NO_3^- levels to prevent vegetation changes and NO_3^- pollution of groundwater, (ii) maximum leaf N contents to prevent disease and frost damage, (iii) minimum pH and maximum inorganic Al concentrations in streams to prevent damage to fish, and (iv) maximum concentrations of inorganic Al and Al/Ca ratios in soil solutions to prevent tree root damage. Empirical data and results of various models using these criteria in the Netherlands show that ambient deposition loads of N and S usually exceed critical loads by factors of two to six. A summary of such critical loads as applied in the Netherlands is given in Table 5–2.

DISCUSSION

The acidification problem in relation to forestry and the way policies have been formulated in Europe to cope with it, should be a useful test for the Montreal Process. This is so for several reasons. First, because acidification is usually of regional or continental scale, any policy implementation is immediately relevant for large areas, normally across national boundaries. Second, the large research effort in acidification will help to increase the chance of finding proper criteria and indicators. A third reason is that the critical loads approach could perhaps be adopted in the Montreal Process, and help to make it more specific; however, before discussing the possibilities to adopt the critical load approach in the Montreal Process, we will discuss some pitfalls and problems of that approach.

The processes of acidification of soils and water are well understood and can be quantified fairly well. The same applies to some of the potential ecological effects as determined under controlled conditions in, e.g., the greenhouse; however, deriving critical loads for field conditions is rather problematic. Critical loads derived from linking critical indicator values to simple steady-state mass balance models, are subject to uncertainties stemming from assumptions about both critical indicator values and the models used (De Vries, 1993). This is specifically true for the critical loads for acidity and N related to effects on forest vitality, such as inhibition of uptake, increased sensitivity for drought, pests and diseases and nutrient imbalances. These critical loads are not as well supported by empirical data as those derived for vegetation changes, NO_3^- leaching to groundwater, depletion of readily available Al and damage to fish populations due to Al toxicity (De Vries et al., 1995a).

Uncertainties in the critical indicator values used, (Table 5–2) greatly influence the calculated critical loads. Critical indicator values are based mainly on laboratory and greenhouse experiments (e.g., with tree seedlings), with limited applicability to field situations. Exceedances of critical loads are thus much bet-

Table 5–2. Average critical loads for acidity and N for forest ecosystems in the Netherlands (after De Vries, 1993).

Compound	Effects	Critical indicator value†	Critical loads: Coniferous forests	Critical loads: Deciduous forests
			mol ha^{-1} yr^{-1}	
Acidity	Root damage	Al < 0.2 mol m^{-3}	1 100‡	1 400‡
	Inhibition of uptake	Al/Ca < 1.0 mol mol^{-1}	1 400‡	1 100‡
	Al depletion	$\Delta al(OH)_3$ = 0 mmol kg^{-1}	1 200‡	1 300‡
	Al pollution	Al < 0.02 mol m^{-3}	500‡	300‡
Nitrogen	Inhibition of uptake	NH_4/K <5 mol mol^{-1}	1250–5000§	
	Nutrient imbalance	N/BC < 1.25 mol mol^{1-1}	1000–1400	1000–1400
	Increased susceptibility to frost and diseases	N <1.8 %	1500–3000¶	
	Vegetation changes	NO_3^- <0.1 mol m^{-3}	500–1400#	800–1400#
	Nitrate pollution	NO_3^- <0.4–0.8 mol m^{-3}	900–1500††	1700–2900††

† Background information on the various critical indicator values is given in De Vries (1993). The sign < is used to indicate that parameter values should stay *below* these critical indicator values to avoid adverse affects. Critical Al and NO_3^- concentrations and critical Al/Ca and NH_4/K ratios related to root damage, inhibition of nutrient uptake and vegetation changes refer to the soil solution. Critical Al and NO_3^- concentrations related to pollution refer to phreatic groundwater. Critical N contents related to an increased risk for frost damage and diseases and critical N/BC ratios relates to nutrient imbalances refer to the foliage.

‡ Derived by a steady-state model. Al pollution refers to phreatic groundwater. For groundwater used for the preparation of drinking water, a critical acid load of 1600 mol ha^{-1} yr^{-1} was derived (De Vries, 1993).

§ Derived by a steady-state model assuming no nitrification (first value; worst case) and 50% nitrification in the mineral topsoil (second value).

¶ Based on empirical data on the relation between N deposition and foliar N contents (De Vries, 1993).

The first value is derived by a steady-state model (worst case) and the second value is based on empirical data.

†† Derived by a steady-state model using a critical NO_3^- concentration of 0.4 and 0.8 mol m^{-3} respectively. NO_3^- pollution refers to phreatic ground water. For deep groundwater, the critical load will be higher because of denitrification.

ter correlated with the occurrence of vegetation changes (e.g., Bobbink et al., 1995) and damage to fish populations (e.g., Hendriksen & Hesthagen, 1993) than with forest damage even though there are indications for such relationships. (Lenz & Schall, 1991 and Nelleman & Frogner, 1994, for forests in Germany and Norway, respectively).

The use of a simple mass balance model for the soil is questionable in relation to effects of both nutrient stress and increased natural stress. The development of multi-stress models, including effects of drought, pests, and diseases, are necessary to support the results of such simplified models. The simple mass balance models probably do provide the correct deposition values for target indicator values, as indicated by a first comparison of critical loads for a Norway spruce stand in Solling (Germany), derived with integrated forest- soil models and simple mass balance models (De Vries et al., 1995b).

A more general problem with the critical acid loads approach is that the causal relationships between indicator values and acidification are extremely complex. Many of the symptoms of effects of acidification (indicated by needle yellowing, changes in species composition) are far from specific and may have other causes. Vice versa, even excessive soil acidification (indicated, e.g., by the soil solution composition) often hardly affects tree growth and appearance, even though experiments performed under controlled conditions would led us to expect otherwise. The fact that most European forests grow well in spite of actual atmospheric deposition loads being several times higher than critical loads, also raises serious questions about the approach; however, that fact does not necessarily imply that the critical loads approach is wrong in principle or in practice. First, the time horizon of the present research is probably too short to obtain results that are conclusive on the time scale of forest ecosystems. If indeed uptake of essential base cations by trees to be harvested and removed exceeds weathering rates, the system cannot be sustainable: to sustain production, eventually deposition will have to be decreased or base cations will have to be supplied as fertilizer. Second, negative effects have been established for other above-critical loads for other targets such as aquatic life in forests streams, birds, ground vegetation, and mycorrhizal fruit bodies. Ultimately the public, not the forest manager alone, should decide what ecological changes are acceptable or not.

Summarizing, one may conclude that the uncertainty in critical N loads related to biological diversity that are supported by empirical data, is relatively small, whereas the critical loads for acidity and N related to forest damage mainly have a signal function. An exceedance of these critical loads does not necessarily imply visible effects or even the dieback of forests, but the risk increases with the degree to which present loads exceed critical loads and as excessive inputs persist for a longer time.

Even though the reliability of the critical load approach may be sometimes only apparent, it is certainly more appealing to most managers and policy makers than the vagueness of the criteria and indicators used in the Montreal Process (Anonymous, 1995). This is in part because the Montreal Process deals with a wide range of causes of forest ecosystem disturbance, which often interfere, making it difficult or impossible to identify proper policy measures. On the other hand, the critical loads approach shows that at least for some of the factors causing disturbance, more precise criteria are possible, and should, in our opinion, be incorporated in the Montreal Process.

One aspect of the Montreal Process is of particular potential importance for both sustainable forest management in general and for evaluating the critical loads approach: the availability of monitoring data relevant to the different criteria. The variety and number of (necessarily long-term) data to properly judge changes in the various indicators is probably staggering, and research to help to trim the set of data to the essential minimum should have the highest priority.

REFERENCES

Aber, J.D., K.J. Nadelhoffer, P. Steudler, and J.M. Melillo. 1989. Nitrogen saturation in northern forest ecosystems. Bioscience 39:378–386.

Anonymous. 1995. The Montreal Process. Criteria and indicators for the conservation and sustainable management of temperate and boreal forests. Canadian For. Serv., Quebec, Canada.

Arnolds, E., 1991. Decline of ectomycorrhizal fungi in Europe. Agric. Ecosys. Environ. 35:209–244.

Arts, G.H.P. 1990. Deterioration of Atlantic soft-water systems and their flora. Ph.D. diss. Catholic Univ. of Nijmegen, the Netherlands.

Billet, M.F., E.A. FitzPatrick, and M.S. Cresser. 1990. Changes in the carbon and nitrogen status of forest soil organic horizons between 1949/50 and 1987. Environ. Poll. 66:67–79.

Bobbink, R., M. Hornung, and J.G.M. Roelofs. 1998. The effect of air-borne nitrogen pollutant on species diversity in natural and semi-natural European vegetation. J. Ecol. (In press).

Boumans, L.J.M., and W. Beltman. 1991. Kwaliteit van het bovenste freatische grondwater in de zandgebieden van Nederland onder bos-en heidevelden. Rapport 724901001. Rijksinstituut voor Volksgezondheid en Milieuhygiëne, Bilthoven, the Netherlands.

Boumans, L.J.M. 1994. Nitraat in het bovenste grondwater onder natuurgebieden op zandgrond in Nederland door atmosferische stikstof depositie. Rapport 712300002. Rijksinstituut voor Volksgezondheid en Milieuhygiëne, Bilthoven, the Netherlands.

Branderud, T.E. 1995. The effects of experimental nitrogen addition on the mycorrhizal fungus flora in an oligotrophic spruce forest in Gårdsjön, Sweden. For. Ecol. Manage. 71:111–122.

De Kam, M., C.M. Versteegen, J. Van den Burg, and D.C. Van der Werf. 1991. Effects of fertilization with ammonium sulphate and potassium sulphate on the development of *Sphaeropsis sapinea* in Corsican pine. Neth. J. Plant Pathol. 5:265–274.

De Visser, P.H.B., W.G. Keltjens, and G.R. Findenegg. 1996. Transpiration and drought resistance of Douglas-fir seedlings exposed to excess ammonium. Trees 10:301–307.

De Vries, W., 1993. Average critical loads for nitrogen and sulfur and its use in acidification abatement policy in the Netherlands. Water Air Soil Pollut. 68:399–434.

De Vries, W., G.J. Reinds, and M. Posch, 1994. Assessment of critical loads and their exceedance on European forests using a one-layer steady-state model. Water Air Soil Pollut. 72:357–394.

De Vries, W., E.E.J.M. Leeters, C.M.A. Hendriks, H.F. van Dobben, J. van den Burg, and L.J.M. Boumans. 1995a. Large scale impacts of acid deposition on forests and forest soils in the Netherlands. Stud. Environ. Sci. 63:261–277.

De Vries, W., M. Posch, T. Oja , H. Van Oene, J. Kros, P. Warfvinge, and P.A. Arp. 1995b. Modelling critical loads for the Solling spruce site. Ecol. Model. 83:283–293.

De Vries, W., J.J.M. Van Grinsven, N. Van Breemen, E.E.J.M. Leeters, and P.C. Jansen. 1995c. Impacts of acid atmospheric deposition on concentrations and fluxes of solutes in acid sandy forest soils in the Netherlands. Geoderma 67:17–43.

Driscoll, C.T., C.L. Wyskowski, C.C. Cosentini, and M.E. Smith. 1987. Processes regulating temporal and longitudinal variations in the chemistry of a low-order woodland stream in the Adirondack region of New York. Biogeochemistry 3:225–241.

Dueck, T.A., F.G. Dorel, R. Ter Horst, and L.J.M. Van der Eerden. 1990. Effects of ammonia, ammonium sulphate and sulphur dioxide on the frost sensitivity of Scots pine (*Pinus sylvestris* L.). Water Air Soil Pollut. 54:35–49.

Falkengren-Grerup, N. Linnermark, and G. Tyler. 1987. Changes in acidity and cation pools of South Swedish soils between 1949 and 1985. Chemosphere 16:2239–2248.

Falkengren-Grerup, U., and H. Eriksson. 1990. Changes in soil, vegetation and forest yield between 1947 and 1988 in beech and oak sites in southern Sweden. For. Ecol. Manage. 38:37–53.

Foy, C.D. 1988. Plant adaptation to acid, aluminum toxic soils. Commun. Soil Sci. Plant Anal. 19:959–987.

Graveland, J., R. Van der Wal, J.H. Van Balen, and A.J. Van Noordwijk. 1994. Poor reproduction in forest passerines from decline of snail abundance on acidified soils. Nature (London) 368:446–448.

Grennfelt, P., and E. Thörnelöf (ed.). 1992. Critical loads for nitrogen. *In* Report from a Workshop, Lökeberg, Sweden. 6–10 Apr. 1992. Nordic Council of Ministers,

Hauhs, M., and R.F. Wright. 1986. Regional pattern of acid deposition and forest decline along a cross section through Europe. Water Air Soil Pollut. 31:463–475.

Havas, M., and B.O. Rosseland. 1995. Response of zooplankton, benthos and fish to acidification: An overview. Water Air Soil Pollut. 85:51–62.

Hendriks, C.M.A., W. De Vries, and J. Van den Burg, 1994. Effects of acid deposition on 150 forest stands in the Netherlands: 2. Relationship between forest vitality and the chemical composition of the foliage, humus layer and the soil solution. DLO Winand Staring Centre for Integrated Land, Soil and Water Res. Rep. 69.2. DLO Winand Staring Ctr., Wageningen, the Netherlands.

Henriksen, A., and T. Hesthagen. 1993. Critical load exceedance and damage to fish populations. Inst. for Water Res. Rep. 43. Inst. for Water Res., Oslo, Norway.

Hettelingh, J.P., R.J. Downing, and P.A.M. De Smedt. 1991. Mapping critical loads for Europe. Natl. Inst. of Public Health and Environmental Protection, Coordination Centre for Effects, Bilthoven, the Netherlands.

Hornung, M., M.A. Sutton, and R.B. Wilson (ed.). 1995. Mapping and modelling of critical loads for nitrogen. *In* A Workshop Report. Proc. UN-ECE Workshop. 24–26 Oct. 1994. Grange- over-Sands, England. Inst. of Terrestrial Ecol., Bush Estate, Midlothian, England.

Hultberg, H. 1985. Budgets of base cations, chloride, nitrogen, and sulphur in the acid Lake Gårdsjön catchment, W Sweden. p 133–157. *In* F. Andersson and B. Olson (ed.) Ecological Bull. 37. Swedish Res. Councils, Stockholm.

Kauppi, P.E., K. Mielikänen, and K. Kuusela. 1992. Biomass and carbon budgets of European forests, 1971 to 1990. Science (Washington, DC) 256:70–74.

Kuusinen, M., K, Mikkola, and E-L. Jukola-Suonen. 1990 Epiphytic lichens on conifers in the 1960s to 1980s in Finland. p. 397–420. *In* P. Kauppi et al. (ed.) Acidification in Finland. Springer Verlag, Berlin.

Lammel, R., 1984. Endgültige Ergebnisse und bundesweite Kartierung der Waldschadenserhebung. Allg. Forst Z. 39:340–344.

Lenz, R., and P. Schall, 1991. Belastungen in fichtendominierten Waldökosysteme. Risikokarten zu Schlüsselprozessen der neu-artigen Waldschäden. Allg. Forst Z. 46:756–761.

Lükewille, A, and N. van Breemen. 1992. Aluminum precipitates from ground water of an aquifer affected by acid atmospheric deposition in the Senne, Northern Germany. Water Air Soil Pollut. 63:340–344.

Mallkönen, E., J. Derome, and M. Kukkola. 1990. Effects of nitrogen inputs on forest ecosystems. Estimations based on long-term fertilization experiments. p. 325–347. *In* P. Kauppi et al. (ed.) Acidification in Finland. Springer-Verlag, Berlin.

Marschner, H. 1991. Mechanisms of adaptation of plants to acid soils. Plant Soil 134:1–20.

Mulder, J., and A. Stein, 1994. The solubility of aluminum in acidic forest soils: Long-term changes due to acid deposition. Geochim. Cosmochim. Acta 58:85–94.

Nelleman, C., and T. Frogner, 1994. Spatial patterns of spruce defoliation seen in relation to acid deposition, critical loads and natural growth conditions in Norway. Ambio 23:255–259.

Nilsson, J., 1986. Critical loads for nitrogen and sulphur. Rep.11. Nordic Council of Ministers, Copenhagen, Denmark.

Nilsson, J.,and P. Grennfelt (ed.). 1988. Critical loads for sulphur and nitrogen. *In* Report from a Workshop, Skokloster, Sweden. 19–24 Mar. 1988. Miljø Rep. 15. Nordic Council of Ministers, Copenhagen, Denmark.

Olsson, M., K. Rosén, and P.A. Melkerud. 1991. Mapping base cation budgets for Swedish forest soils. *In* Proc. 2nd Int. Symp. on Environ. Geochem., Uppsala, Sweden. March 1991.

Ramakrishna, K., and E.A. Davidson. 1998. Intergovernmental negotiations on criteria and indicators for the management, conservation, and sustainable development of forests: What role for forest soil scientists? p. 1–16. *In* E.A. Davidson et al. (ed.) The contribution of soil science to the development of and implementation of criteria and indicators of sustainable forest management. SSSA Spec. Publ. 53. SSSA, Madison, WI (this publication).

Roberts, T.M., R.A. Skeffington, and L.W. Blank. 1989. Causes of type spruce decline. Forestry 62 (3):179–222.

Schulze, E.-D., 1989. Air pollution and forest decline in a Spruce (*Picea abies*) forest. Science (Washington, DC) 244:776–783.

Schütt, P., and E.B. Cowling. 1983. Waldsterben, a general decline of forests in central Europe: Symptoms, development, and possible causes. Plant Dis. 69:548–558.

Stein, A., and N. van Breemen. 1993. Time series analysis of changes in the soil solution: Evidence for approach to nitrogen saturation in Dutch forests. Agric. Ecosyst. Environ. 47:147–158.

Sverdrup, H., and P. Warfvinge. 1993. The effect of soil acidification on the growth of trees, grass and herbs as expressed by the (Ca+Mg+K)/Al ratio. Rep. in Ecology and Environmental Engineering 1993. Lund Univ., Dep. Chemical Eng., Lund, Sweden.

Tamm, C.O., and L. Hallbäcken. 1988. Changes in soil acidity in two forest areas with different acid deposition: 1920s to 1980s. Ambio 17:56–61.

Tan, K., and W.G. Keltjens. 1990. Effects of aluminum on growth, nutrient uptake, proton efflux and phosphorus assimilation of aluminum-tolerant and -sensitive sorghum (*Sorghum bicolor*) genotypes. p. 397–401. *In* M.L. van Beusichem (ed.) Plant nutrition: Physiology and applications. Kluwer Academic Publ., Dordrecht, the Netherlands.

Termorshuizen, A.J. 1990. Decline of carpophores of mycorrhizal fungi in stands of *Pinus sylvestris*. Ph.D. diss. Agricultural Univ., Wageningen, the Netherlands.

Thimonier, A. J.L. Dupouey, F. Bost, and M. Becker. 1994. Simultaneous eutrophication and acidification of a forest in north-east France. New Phytol. 126:533–539.

Tyler, G. 1987. Probable effects of soil acidification and nitrogen deposition on the floristic composition of oak (*Quercus robur* L.) forest. Flora (Jena) 179:165–170.

Ulrich, B., and E. Matzner. 1983. Abiotische Folgewirkungen der weitraümigen Ausbreitung von Luftverunreinigung. Umweltforschungsplan der Bunderministeriums des Innern. Forschungsbericht 10402615, Germany.

Van Der Eerden, L.J. ., W. de Vries, P.H.B. Visser, H.F. van Dobben, E.G. Steingröver, Th.A. Dueck, J.J.M. van Grinsven, G.M.J. Mohren, A.W. Boxman, J.G.M. Roelofs, and J. Graveland. 1995. Effecten op bosecosystemen. p. 63–114. *In* G.J. Heij and T. Schneider (ed.) Eindrapport Additioneel Programma Verzuringsonderzoek, derde fase (1991–1994). Rep 300-05. Rijksinstituut voor Volksgezondheid en Milieuhygiëne, Bilthoven, the Netherlands.

Van Dobben, H.F. 1993. Vegetation as a monitor for deposition of nitrogen and acidity. Ph.D. diss. Univ. of Utrecht, the Netherlands.

Van Dobben, H.F., C.J.M. ter Braak, and G.M. Dirkse. 1999. Undergrowth as a biomonitor for deposition of nitrogen and acidity in pine forest. For. Ecol. Manage. (In press).

6 Role of Criteria and Indicators for the Sustainable Management of Russian Forests

Vladimir N. Gorbachev
Krasnoyarsk Agricultural University

Andrei P. Laletin
Sukachev Forest Institute
Siberian Branch of the Russian Academy of Sciences

The extent of Russian forests renders their global importance unquestionable. The total forest area in Russia is 1.181 billion ha, and 94% of this area (1.111 billion ha) is managed by the Federal Forest Service of Russia. According to Russian forest legislation, the main forest organizing unit is *leskhoz*—a Russian word for the forest management unit (FMU). There are >2000 FMUs in Russia, which are the local units for forest management. Their size varies from 4000 ha in European Russia to >10 million ha in Siberia.

The Russian Federation is the only country-member of both Helsinki and Montreal Processes. As a result of these two processes, the following list of six criteria and 40 corresponding indicators have been proposed at the national level in Russia: (i) maintenance and conservation of productive capacity of forests; (ii) maintenance of acceptable health and vitality of forests; (iii) maintenance and conservation of protective functions of forests; (iv) conservation and maintenance of biological diversity of forests and forest contribution to global C cycle; (v) maintenance of socio-economic functions of forests; and (vi) instruments of forest policy for forest conservation and sustainable management.

Sustainable management of Russian forests offers long-term socially, economically, and ecologically mutually-profitable relationships. Use of forest resources must not result in disappearance or degradation of exploited forests. Sustainable forest management (SFM) also implies the conservation of forests as a part of Russian landscapes. Maintenance of biodiversity and forest productivity at a level acceptable for forest ecosystems represents an invariable basis of SFM. Management of Russian forests shall be realized on the basis of scientific knowledge, experience and a comprehensive assessment of possible impacts on forest ecosystems. SFM implies a multipurpose, uninterrupted and sustained use

The Contribution of Soil Science to the Development of and Implementation of Criteria and Indicators of Sustainable Forest Management. SSSA Special Publication no. 53.

of forest resources, functions, and characteristics, both commercial (timber, fruits, and others) and noncommercial (i.e., preservation of historic traditions).

Prior to the eighth meeting of the Montreal Process held in June, 1996, in Canberra, Australia, the member countries completed a questionnaire designed to assess data availability and reliability, and the capacity to report on each indicator. This information was compiled in 1996 by the Montreal Process Liaison Office in Canada into a joint report "Status of Data and Ability to Report on the Montreal Process Criteria and Indicators." That report also identified areas of difficulty, including those indicators each country found most challenging for reporting.

Russia has indicated that it has the ability to report easily on 13 indicators; easy to moderate, 2; easy to difficult, 4; moderate, 11; moderate to difficult, 9; difficult, 7; impossible, 19; and not rated, 2. For forest soil scientists the most interesting is Criterion 4: Conservation and maintenance of soil and water resources. Data on corresponding indicators on this Criterion are presented in Table 6–1. Of the seven Criteria of the Montreal Process, information on Criterion 4 is the most limiting.

CURRENT SITUATION WITH THE SOIL COVER IN SIBERIAN FORESTS

In our opinion, the current situation with respect to the soil cover in forested areas of Russia is unfavorable. Management activity in the forests is leading to the development of many negative processes, such as erosion, compaction, disturbance of their structural condition, loss of humus and nutrients, water logging, and radioactive contamination.

There are no complete data on the condition of Siberian forest soils; however, in the territory of Krasnoyarsk region alone, there are >11 million ha of land that have been eroded or are vulnerable to erosion (including agricultural lands), 3 million ha of acid soils, and 228 million ha of poorly drained soils (Idemechev, 1996). Contamination of the soils and vegetation by Pu 239 and 240 within the limits of the sanitary-protective area of the Krasnoyarsk mining-chemical enterprise is of great concern. Radioactive contamination, not only in the soils of the radioactive contamination zone, but also on the sites that have been chosen for comparison, is tens of times more than global background levels.

A soil cover study of the vast territory of Siberia, where forest industry is one of the priorities, has shown examples of elimination of the forests and soil cover. Soil loss is most obvious where logging occurs on light textured soils (>85% sand). After logging in pine forests on light soils and further ploughing of the soils, the following forms of elements are lost from the humus horizon: 55 to 75% of C, 10 to 15% of easily hydrolyzed N, 20 to 30% of NH_4–N, 25 to 80% of NO_3–N, 30 to 60% of exchangeable Ca, and 65 to 75% of exchangeable Mg (Babintseva et al., 1984; Gorbachev et al., 1991; Gorbachev & Popova, 1992).

Of particular concern is logging in mountain forests on soils with low resistance to erosion due to steep slopes and sandy texture. The surface gets heavily

Table 6–1. Assessment of Russian forest data for Criterion 4 of the Montreal Protocol.

		Data quality							
Indicator	Data availability	Reliability	Historical coverage	Frequency of collection	Source of aggreg ation	Area of coverage	Future data availability	Reporting methodology	Difficulty of reporting
Area with erosion	Some	Reliable	since 1966	5 years	Nationally, Ecological zones	100%	High	Quantifiable	Easy to difficult
Area managed for protection	Some	Reliable	since 1966	5 years	Nationally, Ecological zones	100%	High	Quantifiable	Easy to moderate
Area with altered stream flow	Not presently						Medium	Proxy	Impossible
Area with depleted organic matter	Not presently						Medium	Proxy	Impossible
Area with compaction	Not presently						Medium	Proxy	Impossible
Area with altered biotic diversity	Not presently						Medium	Proxy	Impossible
Water bodies with altered chemistry	Not presently				Ecological zones		Low	Proxy	Impossible
Area with accumulation of toxins	Some	Reliable	No data	No data	Nationally		High	Proxy	Impossible

eroded in the logged areas, infiltration of the soils falls sharply, and surface flow increases. After forest logging, infiltration drops from 40 to 75 mm min^{-1} to 1 mm min^{-1}, and erosion losses can reach up to 10 to 28 thousand tons km^{-2}, depending on the steepness and exposure of the slopes (Krasnoshekov & Gorbachev, 1987); however, even in the conditions of relatively flat relief of the Yenisey river zone in Siberia, forest logging leads to heavy erosion of the soil surface (Babintseva et al., 1984). The area of a logging unit that gets eroded is 80 to 85% (60% of the area heavily eroded) when heavy machinery is used. Forest litter and the upper horizons of the soil can be completely eliminated; compactness of the soils increases sharply (by 1.5 to 2 times), and total soil porosity and especially air-filled porosity decrease (by 3 to 4 times). Simultaneously, the mineral nutrients of plants change considerably—C, N, and P content decrease by one-half. The number of microorganisms, especially in the areas where machinery operates, decreases by 10 to 15 times. Populations of ammonifying, N fixing, and cellulose decomposing bacteria are reduced, leading to sharp declines in decomposition and intensity of CO_2 emission (Babintseva et al., 1984). Thus, many negative factors are present that may cause sharp declines in soil productivity.

Fires play a particular role in formation of forest ecosystems. They cause change in tree species composition, influence the age structure of the stands, and disturb and change the composition of the grass–bush layer. The spatial distribution of trees of different age groups and generations and the heterogeneity and mosaic structure of the soil cover are related to a considerable degree with fire. First, high intensity fires in the regions where frozen and seasonally frozen soils are widespread, and where tree species have a surface root system, help to form the microrelief pattern of hillocks and pits (the result of post-fire windfall). An integral part of the microrelief are fallen tree trunks that decompose slowly. Formation of such relief forms leads to disturbance of the soils and causes mixing of different genetic soil horizons. The soils that vary within this microrelief are distinguished by water-temperature regime, humus, nutrients, exchangeable cation content, and indicators of biological activity (Gorbachev & Popova, 1996).

The quantity of ash from the burning of fallen trunks can reach up to 3.8 to 9.9 kg m^{-2}. The acidity of litter within 2 wk after fire is different on the sites with different fire intensity. Pyrolysis of organic matter causes a shift of acidity towards neutralization. With a high intensity fire, the pH of water extracts of ash can reach up to 7.4 to 9.4. The change of acidity of the upper soil horizons is mostly noticeable the year of the fire or the following year. With a high intensity of fire and the burning of humus, the content of vegetation nutrient elements and exchangeable bases is sharply reduced (Gorbachev et al., 1982).

Fire influences significantly the biological condition of the soil. High intensity fires sharply depress soil biological processes. The soil microbial complexes change considerably: mycelia of fungi and actinomycetes practically disappear; spore forms of microorganisms and bacteria predominate; the number of ammonificators and cellulose decomposers decreases; and microoganisms use mineral rather than organic N.

It has been estimated that 90% of forest fires in Russia are caused by human activities (Sofronov & Vakurov, 1981). Given the large impacts of fire on soil processes in these ecosystems, the incidence and intensity of fires should be included as an indicator of human effects on forests and forest soils.

SUGGESTIONS AND COMMENTS ON CRITERIA AND INDICATORS FOR SUSTAINABLE FOREST MANAGEMENT FROM THE SOIL SCIENTIST POINT OF VIEW

A complete inventory of forest resources is necessary (timber stock, soil resources, soil cover, and others). In this respect, cartography of the soil cover and its condition is needed to develop a system of sustainable forest management (Gorbachev et al., 1989, 1993). One soil map is inadequate to solve the many problems of sustainable forest management. It is desirable to compile a series of soil–ecological maps of: stocks of humus, N, exchangeable bases, acidity, soil cover disturbance, forest-vegetation properties of the soils, and others. Soil-ecological maps can serve as the basis for various forest management and forest industry goals with respect to sustainable forest management. These include the elaboration of forest logging methods, elaboration of methods to assist natural regeneration, and, in the areas with poor regeneration, to select tree species in accord with forest-vegetation properties of the soils. Soil-ecological maps also can be used to determine appropriate steps to avoid possible water logging of cut areas, to determine measures to drain water-logged forest soils, to organize the territory, and to choose sites for agricultural use for the creation of a forage and food-producing base in the area.

The natural basis of sustainable forest management in a specific area should first be ecological zoning of the territory. It can be performed according to water-collecting river basins (in the plains) or according to altitude belts (in the mountains).

Some of the indicators were identified by us before the elaboration of C&I of the Montreal Process within the framework of the international project "Alternative Forestry in the Murma River Basin" (Babintseva & Gorbachev, 1994, 1996). A great variety of the soil covers were revealed in the Murma river basin in the course of the project work. Soil cover, along with other factors, served as a basis for defining ecological zones: ecologically unstable territories (water-logged parts, areas with fine stony soils, areas where gentle forest use techniques are required; parts where the soils are light according to granulametric composition, steep parts of the slopes etc.); protective zones near waterflows; and sustainable forest use zones. A forest management system was developed for each of the ecological zones. Its focus was on the preservation of the forest's basic functions (resource, water protection, soil protection, and others). Elaboration of measures within the systems of forest management for the ecological zones was based on the use of large scale maps of forests, soils, and the zones of ecological disturbance and others. Hence, many of the proposed criteria and indicators were already in use in 1993 to 1994 in this case study.

With respect to the criteria proposed for the Russian Federation, we recommend the following:

Criterion 1. Maintenance and preservation of the forest's productive capacity. This criterion will be effective only if all forest exploitation occurs on an ecological basis. To do so, it is necessary in each case first to establish ecological zoning of the lands, i.e., to determine the territories in which forest use is possible at the present or in the future. It is necessary to have good information on natural peculiarities of the territory. It is extremely important to have as complete information as possible on soils and their forest-vegetation properties, because maintenance and preservation of the forest's productive capacity is directly dependent on the preservation of soils.

Criterion 2. Maintenance of acceptable sanitary condition and viability of forests. Indicator 3b of the Montreal Process needs to be included in this criterion: area and percentage of the forests characterized by reduction of biological components that indicate changes in the fundamental ecological processes (i.e., such as condition of the soil, nutrient cycles, and others).

Criterion 3. Preservation and maintenance of the forest's protective functions. It is necessary to indicate not only a quota for the forest land area that is to be used to protect the soils, as is proposed in the Russian Federation criteria, but also that the whole area of forest soils is subject to the criteria and indicators, as is proposed in the Montreal Process. Safety of the soil cover over all Russia will not only preserve and maintain the protective functions of the forests, but also preserve the forest as an ecosystem. For this criterion, it is necessary to define land areas covered with forest that have a long frost period. Forest logging by modern methods on such soils leads to unexpected consequences (sharp development of solifluctional processes, water erosion, broad development of the surface hard flow, and others).

Criterion 4. Preservation and maintenance of the forest's diversity and their contribution into the global C cycle. It is necessary to add the N cycle.

As short-term indicators of forest soil conditions, the following are suggested: compaction, pore volume of aeration, biological activity, qualitative and quantitative composition of microflora, humus, and N stocks in the soils. These indicators should be measured once every 10 yr in undisturbed forests and once every 5 yr in logged forests. Soil–ecological cartography should be performed once every 10 yr.

ACKNOWLEDGMENTS

Support for preparation of this chapter was provided by grants from the W. Alton Jones Foundation and the Trust for Mutual Understanding for the Friends of Siberian Forests.

REFERENCES

Babintseva, R.M., V.N. Gorbachev, and N.D. Sorokin. 1984. Ecological aspects of reforestation with modern logging. Russian J. For. Sci. 5:19–25 (In Russian).

Babintseva, R.M., and V.N. Gorbachev. 1994. Forest management ecosystem planning in large rivers basins. p. 111–124. *In* R.G. Khlebopros (ed.) Global and regional ecological problems. Krasnoyarsk Publ., Krasnoyarsk, Russia.

Babintseva, R.M., and V.N. Gorbachev. 1996. The Murma river project: Developing sustainable forestry in Siberia. *Taiga News* 16:9–10.

Gorbachev, V.N, R.M. Bavintseva, N.D. Sorokin, and E.P. Popova. 1991. Evaluation of forest soils condition in fresh felled areas. p. 112–117. *In* S.V. Zonn (ed.) Degradation and restoration of forest soils. Nauka Publ., Moscow, Russia (In Russian).

Gorbachev, V.N., V.D. Dmitrienko, E.P. Popova, and N.D. Sorokin. 1982. Soil–ecological investigations in forest ecosystems. Nauka Publ., Novosibirsk, Russia (In Russian).

Gorbachev, V.N., V.M. Korsunov, and T.B. Baranchikova. 1989. Structure of the soil cover in the south of Siberia. Geography Nat. Resour. 3:74–82 (In Russian).

Gorbachev, V.N., V.M. Korsunov, and T.B. Baranchikova. 1993. Experience on cartography of the soil cover structure in taiga landscapes using aerospace information. p. 150–161. *In* B.P. Tchesnokov and A.V. Chuchalin (ed.) Scientific-technical collection on geodesy, aerospace survey and cartography. Experience on complex cartography and study of the environment in Krasnoyarsk territory based on the materials of aerospace survey. Central Scientific Res. Inst. of Geodesy, Aerospace Survey and Cartography Spec. Publ., Inst of Geodesy, Aerospace Survey and Cartography, Moscow, Russia (In Russian).

Gorbachev, V.N., and E.P. Popova. 1992. Soil cover of the southern Taiga in middle Siberia. Nauka Publ., Novosibirsk, Russia (In Russian).

Gorbachev, V.N., and E.P. Popova. 1996. Fires and soil formation. p. 331–336. *In* I.G. Goldammer, and V.V. Furyaev (ed.) Fire in ecosystems of Boreal Eurasia. Kluwer Academic Publ., Dordreckt, the Netherlands.

Idemechev, V.F.. 1996. On condition of the environment in Krasnoyarsk territory report, 1995. Krasnoyarsk Regional Ecological Committee, Krasnoyarsk, Russia (In Russian).

Krasnoshekov, Ju.N., and V.N. Gorbachev. 1987. Forest soils of the lake Baikal basin. Nauka Publ., Novosibirsk, Russia (In Russian).

Sofronov, M.A., and A.D. Vakurov. 1981. Fires in the forest. Nauka Publ., Novosibirsk, Russia (In Russian).

7 Utility of Montreal Process Indicators for Soil Conservation in Native Forests and Plantations in Australia and New Zealand

C. T. Smith

New Zealand Forest Research Institute
Rotorua, New Zealand

R. J. Raison

CSIRO Forestry and Forest Products
Canberra, Australia

The Santiago Declaration, signed in February 1996, agreed on a set of Criteria and Indicators (C&I) for the sustainable management of temperate and boreal forests. This outcome was a result of the Montreal Process (Anonymous, 1995) involving input from 10 countries, including Australia and New Zealand, responsible for management of about 60% of the world's forests. Seven criteria were defined, with one dealing with the conservation and maintenance of soil and water resources; and within this, eight possible indicators were proposed, four of which deal specifically with the soil (see Ramakrishna and Davidson, 1998, this publication).

Whilst the focus of the Montreal C&I is more at the national level, it is clear that soil must be protected at the local forest management unit (coupe or compartment at the scale of hectares) level, and that aggregation of data at higher (e.g., regional) levels will only be meaningful if based on accurate information collected at the local level. Thus indicators and monitoring for soils must be applicable to the local scale.

The role of C&I in achieving sustainable forest management is shown in Fig. 7–1. Codes of forest practice are statements of goals and guidelines for achieving environmental care; these are implemented on the ground via local management prescriptions. C&I provide the tools for monitoring the consequences (outcomes) of management, and environmental standards assist interpretation of measured trends. Findings may lead to changes in codes or prescriptions (i.e., lead to adaptive management). For soils, monitoring is needed on representative managed coupes–compartments.

In this chapter, we evaluate the utility of the proposed Montreal Indicators for soil conservation for demonstrating sustainable forest management in

The Contribution of Soil Science to the Development of and Implementation of Criteria and Indicators of Sustainable Forest Management. SSSA Special Publication no. 53.

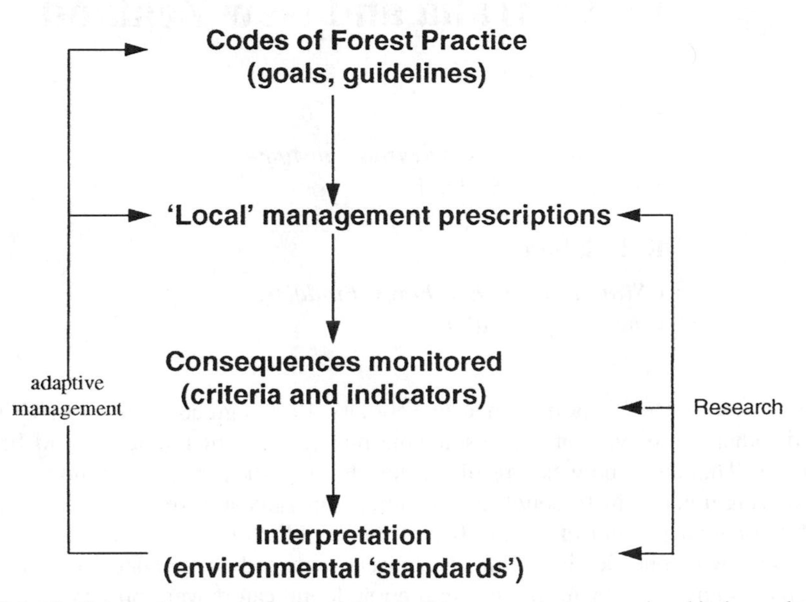

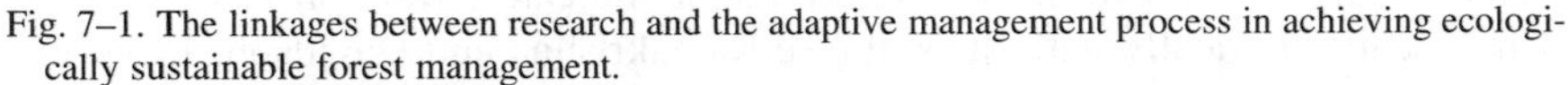
Fig. 7–1. The linkages between research and the adaptive management process in achieving ecologically sustainable forest management.

Australia and New Zealand. The approach to adopting these indicators in the two countries is briefly outlined. Issues addressed include: consequences of differing management objectives in native forests and plantations; the scientific basis for indicators; monitoring considerations; interpretation of trends; and ways to progress the role of indicators in achieving sustainable forest management.

THE FORESTRY SECTOR AND POLICIES RELATED TO SUSTAINABLE FOREST MANAGEMENT

Australia

Wood production in Australia is based on native [mostly eucalypt (*Eucalyptus* sp.)] forests, and on plantations [mostly radiata pine (*Pinus radiata* D. Don) and sub-tropical pines, but increasingly eucalypts]. There are about 41 million ha of native forests, but only about 7 million ha is currently available and suitable for wood production. Significant areas within the 7 million ha of native forest designated for wood production are managed primarily for conservation, and this area is likely to increase during the implementation of a national reserve system during the next 1 to 2 yr. There is a trend toward zoning of the wood production forest into areas to be conserved, areas to be lightly logged where conservation is given priority, a general management zone with a balance between conservation and production, and intensively managed areas for wood production. There are currently about 1 million ha of plantations, of which about 80% are *Pinus radiata*. Eucalypt plantations, mostly for pulpwood, are increasing, but total (pines plus eucalypt) rates of new planting are only about 25 000 ha yr^{-1}. This is expected to increase, and there is an industry–government vision to treble the size of the plantation estate within 30 yr. Most of the expansion will be on farm land.

Following several decades of intense community debate about forest use, and several major inquiries (e.g., Resource Assessment Commission, 1992), the Federal and State Governments have signed a National Forest Policy Statement (NFPS, 1992) that established broad guidelines for the future management of Australia's forests. The NFPS is based on the following key thrusts: (i) establishment of a comprehensive, adequate, and representative (CAR) national reserve system; (ii) management of all forests on an ecologically sustainable basis; and (iii) defining conservation, production, and socio-economic objectives.

Sustainable forest management is to be achieved in each State via Codes of Forest Practice that will eventually include baseline environmental standards. A Technical Working Group on Forest Use and Management has attempted to define such standards (TWGFUM, 1995). The working group concluded that at this time it is only possible to provide goals and guidelines for sustainable management, but that further research and the development of a national set of monitoring protocols should be directed toward the development of national environmental standards.

For soil protection, the goals and guidelines proposed by the TWGFUM may form the basis for an initial set of indicators to be evaluated in native forests. This evaluation will be conducted in a series of case studies in important representative forest types, and examine the relationships among the indicators and important ecosystem processes (e.g., nutrient and water availability, growth), as well as the cost-effectiveness of monitoring the indicators. It must be emphasized that single factor relationships among proposed indicators (e.g., soil organic matter) and growth are not to be expected; usually interactions between several factors control tree growth. The following provisional goals and guidelines were proposed for soils: (i) the area of harvested coupes–compartments occupied by zones of major soil disturbance (log dumps, access tracks, snig tracks with rutting of >10 cm in depth) should be minimized and confined to ≤20%; (ii) for the harvested portion of coupes–compartments, but excluding log dumps and truck access tracks, soil porosity should be maintained at >90% of the value existing before harvesting; (iii) site (i.e., litter plus soil to 30 cm depth) organic matter and nutrient supplying capacity, measured 3 to 5 yr after harvesting, should be maintained at >85% of the value existing prior to harvesting; (iv) soil erosion should not be accelerated from the coupe–compartment, and adequate precautions should be taken to minimize soil redistribution downslope within the harvested area; and (v) logging practices should be varied according to changes in soil erosion hazard classes.

These indicators show a degree of commonality with those proposed for soil protection under the Montreal Process. Soil physical disturbance, erosion, and organic matter change are captured in both schemes, but for Australia an indicator for the accumulation of persistent toxic substances is not seen as important. The link between soil transport and water quality is well developed in both cases.

A review of the proposed Montreal indicators for soils (FWPRDC, 1996) concluded that, in general, these could not be applied in the near future in Australia. Major impediments were lack of historical data, lack of proposed monitoring by land managers at appropriate scales, and lack of a scientific framework for interpreting trends (e.g., for defining *significant* change). It was considered to be unproductive to commence broad scale monitoring until much more is known about what to measure (i.e., relevance of proposed indicators), and how to conduct monitoring in a cost-effective way. These factors are to be examined as part of the proposed case study approach referred to above.

New Zealand

The total land area of New Zealand is 27 million ha. About 52% (14.1 million ha) is pasture and arable land; 19% (5.0 million ha) is other nonforested land; 24% (6.4 million ha) is natural, indigenous forest (based on the National Forest Survey Data Base, conducted 1945–1953 and updated in 1974); and 5% (1.5 million ha) is planted production forest.

Indigenous forests historically covered 80% of the country; and only 15% of rich, lowland forest types remain. The majority of indigenous forests are currently protected for conservation purposes under a combination of legislation and voluntary measures (Table 7–1). As a result, only 2% (150,000 ha) of indigenous forests are currently available for timber production (NZMOF, 1993).

Plantation forests managed for production purposes are composed of 90% *Pinus radiata*, 5% Douglas-fir [*Pseudotsuga menziesii* (Mirbel) Franco], 3% other exotic softwoods; and 2% all exotic hardwoods (including eucalypts). The amount of land planted with exotic species for conservation and protection (e.g., stream bank stabilization and erosion control) purposes has not been quantified with any precision. Planted production forests supplied 98.7% of the total roundwood removals from New Zealand forests in 1994 (NZMOF, 1994). Production forest exports were projected to account for 13% of total New Zealand exports in 1995 (NZFOA, 1995). The rate of new planting in 1993 was 61 600 ha (NZMOF, 1995b); and production forests are predicted to increase to 4 million ha in the future as marginal pastoral lands are converted to forestry use. The plantation estate is managed relatively intensively, primarily for wood products.

Relatively fast growth rates for *Pinus radiata* and several other exotic plantation species allow New Zealand to achieve its forest products export earnings from 5% of its land base. This pattern of land use allocation is essential for achieving the requirements of the Montreal Process C&I at a national level. Achieving sustainable forest management at the coupe–compartment level will be realized by the implementation of the Resource Management Act (Table 7–1), and through voluntary compliance with the requirements of the Montreal Process

Table 7–1. Legislation and voluntary measures that promote the conservation of New Zealand indigenous forests.

Key Legislation

The Resource Management Act 1991

Introduced to promote the sustainable management of natural and physical resources. Focus is on *effects* of resource management activities, rather than on activities themselves. Includes scope for protection of areas of significant indigenous vegetation.

The Forests Amendment Act 1993

Has the purpose of promoting the sustainable management of indigenous forests.

The Forestry Rights Registration Act 1983

Introduced to facilitate forestry joint ventures, which involve the combining of resources such as land, capital, and labor to make the best use of those resources for the benefit of all participating parties. The landowner grants the investor the right to establish, manage, and harvest a forest on his or her land, with the landowner retaining land ownership and the use of the land compatible with the needs of the forest.

The Conservation Act 1987

Introduced to promote conservation of New Zealand natural and historic resources, and established the Department of Conservation (DOC). DOC is responsible for managing the majority of the conservation forest estate.

Soil Conservation and Rivers Control Act 1959
(Sections 34 and 35 super-ceded by the Resource Management Act)

Introduced to maintain indigenous forest cover on private land where there are potential erosion problems.

Voluntary Measures

The New Zealand Forest Code of Practice

Developed by Logging Industy Research Organization (1993) to provide a self-regulating guide to the forest industry to meet the requirements of sound and practical environmental management through safe and efficient forestry operations. Designed for operations in planted and natural production forests to address sustainability issues such as: commercial values, soil and water values, scenic values, cultural values, recreational values, scientific and ecological values, forest health, site productivity, off-site impacts, and safety.

The New Zealand Forest Accord 1991

A voluntary agreement between New Zealand major forest industry organisations and conservation groups. Agreed that New Zealand forest industry will not clear native forest or regenerating native forests for establishment of plantations of exotic species.

The Forest Heritage Trust Fund 1990

The Forest Heritage Trust Fund was established to protect remaining indigenous forests and associated vegetation for preserving genetic diversity in indigenous flora and fauna, to safeguard heritage values, and to ensure that nature conservation in an integral part of a sustainable land management ethic.

Nga Whenua Rahui 1990

As a companion to FHTF, this encourages Maori owners of indigenous forests to formally protect their forests while maintaining their tikanga (custom, way of doing things) and tino rangatiratanga (exact evidence of breeding and greatness), to manage their forests sustainably, and to protect their spiritual relationship with the forest, as well as use forests for food, medicine, weaving, and carvings.

C&I and other measures developed by the major industrial, governmental, and private components of the New Zealand forestry sector (Table 7–1).

New Zealand has not developed provisional goals and guidelines for soils, as reported for Australia in this chapter. Current legislation and voluntary measures developed for conserving native forests (Table 7–1) will be useful for achieving soil conservation; however, the wording of these documents is subjective, and requires the development of site-specific standards for soil conservation based on an ongoing case study approach, before operational implementation at the coupe–compartment scale is possible.

Forestry research required to develop a site-specific understanding of the relationship between Montreal Process C&I is supported by funding from the New Zealand government Foundation for Research, Science and Technology (FRST), and by the forest industry sector. Research results applicable for evaluating the adequacy of the Montreal Process C&I are discussed below.

EVALUATION OF PROPOSED MONTREAL INDICATORS FOR CONSERVATION AND MAINTENANCE OF SOIL RESOURCES (CRITERION 4)

This criterion is essential for achieving sustainable forest management. Local and global levels of productivity, biodiversity, and atmospheric conditions are directly related to soil and water conditions, as is the ability of forests to meet national environmental, social, and economic objectives.

The Montreal indicators were defined as quantitative or qualitative variables that can be measured or described periodically to identify trends. They provide an objective means of assessing the status of a region, and for determining if temporal change is occurring. Some of the proposed indicators focus too much on retrospective assessment of damage caused by historical forestry operations (e.g., area and percentage of land with significant soil erosion). More focus should be on the development of best management practices that prevent significant damage from occurring; however, we note that other C&I are complementary, e.g., Criterion 7 includes indicators that encourage best practice codes for forest management, and the extent to which a nation has the capacity to conduct and apply research and development aimed at improving forest management and delivery of forest goods and services, including development of scientific understanding of forest ecosystem characteristics and functions. Hence, the full value of Criterion 4 indicators can only be realized by achieving acceptable levels of related C&I.

The Montreal C&I need to fit into an overall framework for achieving sustainable management (Fig. 7–1). Such a conceptual framework is required to enable the forestry sector to apply them. We suggest that the C&I be placed into a comprehensive framework similar to that suggested for achieving environmentally acceptable deployment of bioenergy production systems (Smith, 1995). The suggested approach involves four components: (i) a code of practice for environmentally acceptable forest management based on a combination of theoretical (e.g., computer simulation) and validated concepts for sustainable forest management, (ii) an ongoing research program designed to test hypothetical concepts, provide forest-based validation for ecologically sound practices, and provide the basis for interpretation of trends in indicators, (iii) an ongoing monitoring program to demonstrate compliance with the code of practice, and to provide evidence that ecosystem sustainability is being achieved, and (iv) an information support system based on components of information storage and retrieval, e.g., Hypertext (Nielsen, 1989), expert system (Proe et al., 1994), and geographic information system (GIS). Forest management systems typically lack Component 3, and thus have a restricted capacity for continuous improvement.

The Montreal C&I are compatible with this suggested approach, since (i) the code of practice and (ii) requirement for research and development are included in Criterion 7; (iii) monitoring of the status of managed forests would take place under Criteria 1 through 5; and (iv) development and maintenance of appropriate information systems are covered by Criterion 7. The following evaluation of specific indicators is made assuming that Criteria 1 through 5 are intended to be the elements of a monitoring system (iii, above) that is essential to the overall suggested framework.

All the proposed Montreal indicators for soil and water resource conservation, except Indicator 4b, have been designated "b" indicating measurement or description will "require gathering new or additional data and/or a new program of systematic sampling or basic research." We agree with this assessment; but believe that there also are significant additional factors limiting the utility of some indicators. These are described below.

While most indicators are adequate in concept, defining the significant degree of change for each indicator will be the largest technical difficulty to be overcome before trend analysis can be conducted and used to assess the sustainability of a given forestry operation. Indicators will require local calibration to determine their relationship with important ecosystem processes (e.g., forest growth, level of suspended sediment in water), and hence levels of tolerable change (thresholds). For many of the indicators proposed (e.g., erosion, soil organic matter), it is not possible to generally specify quantitative values for the threshold between sustainable and unsustainable levels. In these cases, it will only be possible to apply the four-component approach suggested above, and operate under the implicit expectation that the code of practice will be revised as new information becomes available.

It is necessary to distinguish between direct and indirect (e.g., inferred from changes in forest growth) measures of soil quality; and to quantify the links between the soil resource and ecosystem health and productivity. Forest productivity is directly affected by soil quality as well as climate, genotype, and management practices such as stocking and weed competition (Powers, et al., 1990; Dyck & Bow, 1992; Fig. 7–2). Therefore, measuring tree growth rates may not be an adequate measure of changes in soil quality, unless other factors are controlled experimentally. The proposed Montreal indicators for conservation and

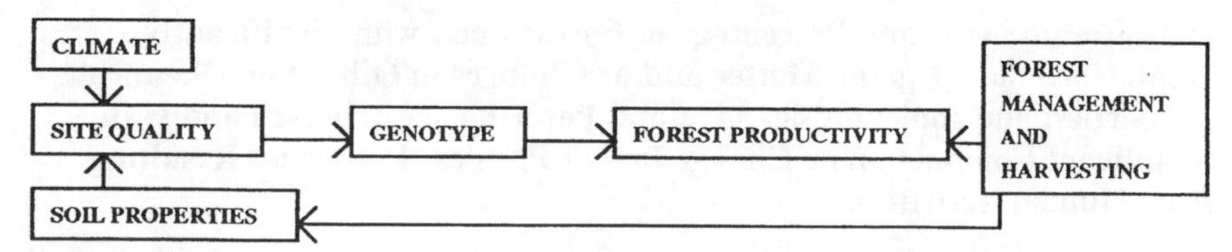

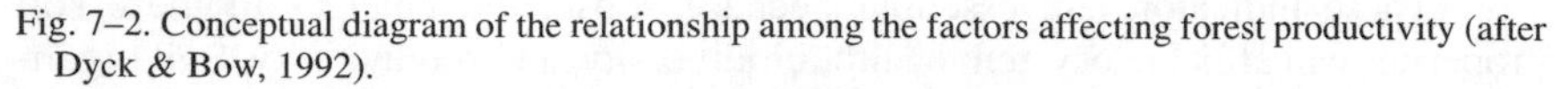
Fig. 7–2. Conceptual diagram of the relationship among the factors affecting forest productivity (after Dyck & Bow, 1992).

maintenance of soil and water resources (Table 1–1 in Ramakrishna and Davidson, 1998, this publication) may be adequate for describing the state of soil resources on some defined spatial and temporal scales, but are not adequate for enabling quantitative inferences about ecosystem health or productivity until locally calibrated via research trials.

Specific Comments on Indicators

The following discussion is limited to indicators (a, d, e, h) associated with soil conservation and the potential for off-site impacts resulting from poor soil management; and are important for forest ecosystem structure, function, and stability (e.g., affecting nutrient and hydrologic cycles).

Indicator 4a. Area and Percentage of Forest Land with Significant Soil Erosion

This indicator is important because eroding landscapes eventually result in lowered ecosystem productivity and water quality, and are not sustainable. This indicator is required to prevent on- and off-site degradation. Studies conducted in New Zealand indicate there are strong relationships among land use practices and associated vegetation, and landscape stability in steep hill country underlain by Tertiary parent materials (Marden & Rowan, 1993). Best management practices have been developed for many forest regions to prevent erosion (e.g., LIRO, 1993); and if well applied can be effective in protecting water quality (Neary & Hornbeck, 1994).

Quantifying soil erosion may be technically difficult over large regions. Studies have identified the difficulty of measuring past erosion damage due to revegetation obscuring erosion scars (Fransen & Brownlie, 1996). These studies suggest that only very severe damage (e.g., gully erosion) or recent scars not obscured by forest cover could easily be measured on vast tracts of forest by aerial survey techniques. Sheet and rill erosion will be difficult to measure accurately. Survey teams would need to distinguish between erosion due to naturally unstable terrain (e.g., steep, young landscapes) and that caused by unsustainable forestry operations. Significant may be defined as erosion greater than occurs naturally, and will have to refer to all types of erosion.

Rather than focus on historical effects, emphasis should be on the future with application of best practice to minimize erosion, monitoring to indicate whether management prescriptions are being effective, and research to enable interpretation of measured changes and to provide management options. Initial effort should concentrate on high risk forest environments.

Indicator 4d. Area and Percentage of Forest Land with Significantly Diminished Soil Organic Matter and/or Changes in Other Soil Chemical Properties; and Indicator 4e. Area and Percentage of Forest Land with Significant Compaction or Change in Soil Physical Properties Resulting from Human Activities.

These indicators are essential because substantial changes in these soil properties can affect ecosystem health, biodiversity, and productivity. Soil organic matter (SOM) has direct relationships with soil biological and physical properties (e.g., bulk density; Federer et al., 1993), water relationships, soil structure, and chemical fertility (e.g., N and P availability, cation-exchange capacity; see Raison & Hopmans, 1995). In addition to affecting soil fertility, SOM and physical properties are important for water quality off-site because soils with high organic matter and low density are generally permeable, and show negligible surface runoff in most cases. Maintaining SOM (C) is important for several Montreal Process criteria, including (i) biodiversity, (ii) ecosystem productive capacity, and (iii) contributions to global C cycles (see Johnson, 1992). Other soil chemical properties contributing to fertility and nutrient dynamics are directly linked to ecosystem biodiversity and productivity.

Whilst SOM is intuitively attractive as an index of soil fertility for the reasons discussed above, its usefulness needs to be established for different forest ecosystems; and a basis for establishing the significance of measured changes established. The relationship between SOM and forest growth is confounded (Fisher, 1995), so that strong single factor relationships between the two should not always be expected. The way in which SOM affects important ecosystem processes (e.g., nutrient supply, infiltration, and storage of water) will vary in different forests, as will their relative importance in contributing to productivity. Hence there is a need for case studies to better evaluate the importance of SOM in representative soils and forests subjected to harvesting.

Defining significant amounts of SOM and other soil chemical and physical properties required for sustainable management will be expensive and requires a thoughtful approach based on case studies in representative stratifications of the forest estate. For extensive forest management systems with low silvicultural inputs, codes of practice must recommend organic matter and nutrient conservation and minimal compaction during harvesting operations. Organic matter and nutrient budgets need to balance over appropriate time scales in such forests. Raison and Khanna (1995) propose that site organic matter (defined as litter plus 0–30-cm depth total organic C, and measured 3–5 yr after harvesting) and soil nutrient supplying capacity be maintained at >85% of the value existing prior to forestry operations. Soil compaction may limit ecosystem productivity for decades or rotations, and will increase risk of surface runoff and erosion. For intensively managed production forests, economic analysis is required to define sustainable management impacts on the soil, assuming the potential for future corrective measures; however, detrimental off-site effects must be prevented.

The difficulty associated with defining significantly diminished soil organic matter or N availability can be illustrated with research data. For example, a trial established on coastal sand dunes in the Woodhill Forest, New Zealand, indicated that significant reductions in *Pinus radiata* growth occurred through plantation age 9 years after removing whole tree and forest floor biomass and associated N equivalent to 48% of that in the ecosystem prior to planting the new crop; but productivity was not reduced following removals of 7 and 17% of total N (Smith et al., 1994a,b); however, on a more fertile site (100 yr old scoria gravel pumice) in the Tarawera Forest, even removals of whole trees (above stump), forest floor and some topsoil did not reduce *Pinus radiata* growth through age 6 yr (Lowe, 1995). Apparently, mineralization of the residual organic N on the recent scoria gravel was adequate to maintain nutrient supply capacity and tree growth in spite of removing approximately 53% of the total N. These results help highlight the difference between the total amount of a nutrient present on site vs. the nutrient supplying capacity of the soil, which is the recommended criterion of Raison and Khanna (1995). For the present, codes of practice can only suggest soil organic matter and nutrient conservation until such time as more calibration studies are conducted in representative physiographic regions. Six such studies are currently under way in New Zealand (Smith, 1994). These trials are located in important production forests across the country; and will be used to define significant reductions in soil organic matter and N in the context of *Pinus radiata* growth and nutrition.

Soil compaction may have positive or negative effects on early (1–3 yr) growth and survival in *Pinus radiata* plantations (McQueen et al., 1994). Compaction of fine textured pumice reduces macro-pore space, and requires ripping to ameliorate soil physical properties to levels required to maintain tree growth and survival; however, compaction of coarse scoria gravel in the Tarawera Forest increased early survival and did not affect height growth of *Pinus radiata* seedlings. Evidently compaction reduced an excessive amount of macro-porosity in the droughty gravel scoria, and improved the characteristics of the rooting zone. Research is being conducted to quantify the relationship between root development and tree growth and survival and soil physical properties on the important soils in the New Zealand plantation estate after machine compaction and amelioration by ripping. This will aid in developing site-specific definitions of significant changes in soil physical properties. Note that the resulting definition will be in the context of *Pinus radiata* requirements, and may not be generic.

Research and observation has shown that natural rates of recovery from soil compaction are very slow, and that physical amelioration (e.g., ripping of log dumps) does not always result in good tree growth. There also is the danger of cumulative effects during a rotation (e.g., from several thinning operations) and between rotations. Selected retrospective studies would be very useful to establish the relationship between initial soil disturbance after logging (assessed, i.e., example, from aerial photographs), current soil physical properties, and accumulated biomass since clearfelling and regeneration. The quantitative effects of soil physical disturbance on long-term growth, and integrated over the scale of the logging coupe, are poorly known. This is a key area for future research.

Although Montreal Indicator 4e focuses on compaction, it should relate to soil disturbance in general, and include soil displacement, inversion and mixing, and puddling. The proportion of a logged coupe receiving different degrees of soil disturbance may be a useful surrogate for soil damage, but this needs to be established for local conditions. Minimizing the spatial area of logged coupes subjected to major soil disturbance (log dumps, access tracks, forwarder tracks with rutting of >10 cm in depth) should be an interim objective until more is known about the impact of soil disturbance on the fertility of forest soils.

The suggested indicators (4d, 4e) propose the area and percentage of forest land with significant soil changes. It will not be feasible to ever establish the area historically affected, although retrospective studies (sampling of replicated sites of known age since harvesting) can be used to improve our understanding of the long-term impacts of past logging and regeneration practices on soil properties. The focus for the future must be on establishing the best parameters to measure, monitoring the effects of current best practice and alternatives, and adapting management practices where necessary.

Indicator 4h. Area and Percentage of Forest Land Experiencing an Accumulation of Persistent Toxic Substances

This indicator is essential to forest sustainability, although refinement will require specific definitions of persistent toxic substances. This phrase is highly politically charged, and subject to confusion because of the failure to distinguish between naturally occurring toxic compounds (e.g., soil Al and Mn; resin acids under pine) and anthropogenic sources (pesticides, wood preservatives, air pollutants). Monitoring for even a limited number of compounds would be very expensive across large tracts of forest. Regulations established for environmentally acceptable pesticide use should be sufficient to prevent toxic accumulations of pesticides from occurring. Although forest managers cannot be held responsible for inputs of toxic compounds associated with air pollution, forests do effectively trap air-borne pollutants, and these can adversely affect forest ecosystems and affect sustainability. Thus their impacts need to be monitored in polluted environments. For example, balanced input-output budgets for base cations in ecosystems receiving acidic deposition requires forest management to evaluate the effects of soil amendments (e.g., harvest residues, wood ash recycling). Forests in Australia and New Zealand are little affected by air pollution. Application of sewage sludge and other effluents to forest sites is increasing and effects on soil and drainage water needs to be monitored.

DISCUSSION

The Santiago Declaration provides a framework for assessing sustainability at national, rather than management unit (coupe–compartment) levels. The distinction between national and coupe–compartment scales is important because it permits coupes to be managed for selected objectives (e.g., water, timber, biodiversity). Separating lands by management objectives may have significant practical and economic advantages; and also provides an opportunity to define the amount of each attribute desired nationally. For example, a nation might ask how the whole forestry sector could be managed to achieve some level of national export earnings or satisfy internal demand for wood products; or what amount of land should be managed for conservation of biodiversity across the range of physiographic zones; however, all forests must be managed sustainably, and for plantations this relates mainly to maintaining the capacity of the soil (together with soil amendment) to support tree growth, and preventing adverse off-site effects on water yield and quality. For native forests, a wider range of values need to be maintained, but not on every hectare of forest at all points in time; however, soil values need to be conserved at the coupe–compartment scale.

In New Zealand, 98% of the remaining indigenous forests are managed primarily for conservation purposes; and the 2% of indigenous forests managed for timber products must be managed sustainably according to an established code of practice (NZMOF, 1993). Essentially all timber harvested from New Zealand forests in 1994 was from plantations (NZMOF, 1994), and this percentage will probably not change in the future as a result of legislation during the past decade or so (NZMOF, 1995a). This land use allocation decision has simplified matters for the forestry sector, because it is easier to distinguish among management priorities. The primarily exotic plantations are managed intensively for timber production, even though this land use practice must be compatible with the requirements of the Resource Management Act, which requires mitigation of negative environmental, economic, and social effects (Table 7–1). Multiple use management goals *per se* are not required for the exotic plantation estate. This rationalization of forest land use and management priorities among separate portions of the nation's forests might apply to some other timber producing nations, and could help in achieving the requirements of the Montreal Process.

If one accepts that lands managed intensively for wood production purposes may require substantial silvicultural inputs (e.g., fertilizer, pesticides, site preparation, and planting), then it is possible to analyze trends of various indicators in a different context than for extensively managed forests. For example, extensively managed forests with low levels of inputs may require harvest usage standards designed to achieve balanced site nutrient budgets over the long run, taking into account N fixation and atmospheric and weathering inputs. On the other hand, harvest usage recommendations for intensively managed forests could permit high nutrient removals with replacement by fertilizer or other (e.g., legume) inputs. Decisions about the degree of organic matter and nutrient removal, soil compaction, and other measures of soil quality could be based on economic analysis of the relative costs of modifying harvesting equipment and delimbing operations, replacement of organic residues, fertilizer, and the degree of growth loss. Such cost-benefit analysis would answer the question: Is the cost of retaining nutrients and organic matter during harvest greater than the cost of ameliorating soil fertility and correcting growth loss with fertilizer or other soil amendments? Scientists need to be able to advise forest managers whether the adage *an ounce of prevention is worth a pound of cure* applies to harvest removals of nutrients and organic matter and other measures of soil-site quality.

A clear zoning of forest uses, ranging from intensively managed wood production forests, to forests managed more extensively for multiple use, and to conservation forests, may facilitate achievement of sustainable management at both local and national scales. For forests managed primarily for conservation values, with little or no timber extraction, a wide range of sustainability indicators must be considered, as well as ways of addressing trade-offs among output priorities. Conservation forests would require emphasis on monitoring of biodiversity indicators and other indicators covering broad community objectives. Monitoring the effects of management is equally important as in wood production forests. Regions experiencing significant decline due to pests [e.g., brush-tailed possum (*Trichosurus vulpecula* Kerr) introduced to New Zealand from Australia], fire or pollution would require careful monitoring to identify trends in conditions that can not easily be monitored remotely (e.g., biodiversity, C flux). Although soil change is likely to be less marked in conservation than in wood production forests, it still needs to be monitored (e.g., erosion following prescribed burning).

For intensively managed wood production forests, the emphasis would be on preventing off-site environmental problems and on quantifying the sustainability of alternative crop production strategies. Monitoring would likely be more frequent and expensive than that on conservation lands because of greater potential for rapid, significant change. Trends towards certification of forest management as sustainable, using environmental management systems approaches (e.g., ISO 14001) or the Forest Stewardship Council principles, will help achieve environmental monitoring on a high proportion of intensively managed forests. The costs of certification and related monitoring will be borne by society through increased forest product prices. National reporting systems could be designed to utilize data collected as part of voluntary certification initiatives for national reporting requirements of the Montreal Process.

CONCLUSIONS

The proposed indicators which characterize Montreal Process Criterion 4 are in theory adequate to conserve soil resources, and assist with the protective and productive functions of forests; however, most require new data, new sampling systems or basic research, and are defined rather subjectively. In most cases they cannot be applied operationally at present. Acceptable limits of change for indicators need to be defined locally. Indicators of severe disturbance are easier to define; but limits for acceptable soil biological, chemical and physical properties require research to establish a relationship with important ecosystem processes and forest productivity.

Australia and New Zealand are continuing research required to improve forest management and to enable the Montreal C&I to be applied in a range of forest ecosystems. Substantially increased research support is needed to expedite the essential development of indicators of soil fertility applicable at the management unit scale.

ACKNOWLEDGMENTS

The authors thank Eric Davidson for organizing the symposium "Criteria and Indicators for the Management, Conservation, and Sustainable Development of Forests" in conjunction with the 1995 annual meetings of the Agronomy Society of America in St. Louis, MO; C.T. Smith thanks the USDA Forest Service for financial support. Support for research conducted by the New Zealand Forest Research Institute was provided by the Foundation for Research, Science and Technology and the forest industry sector. CSIRO research was supported by the Australian Government and the FWPRDC.

REFERENCES

Anonymous. 1995. Criteria and indicators for the conservation and sustainable management of temperate and boreal forests. The Montreal Process. Canadian For. Serv., Nat. Resour. Canada, Hull, Quebec, Canada.

Dyck, W.J., and C.A. Bow. 1992. Environmental impacts of harvesting. Biomass Bioenergy 2:173–191.

Federer, C.A., D.E. Turcotte, and C.T. Smith. 1993. The organic fraction: Bulk density relationship and the expression of nutrient content in forest soils. Can J. For. Res. 23:1026–1032.

Fisher, R.F. 1995. Soil organic matter: Clue or conundrum. p. 1–11. *In* W.W. McFee and J.M. Kelly (ed.) Carbon forms and functions in forest soils. Proc. 8th North America Forest Soils Conf., Gainesville, FL. May 1993. SSSA, Madison, WI.

Fransen, P., and R. Brownlie. 1996. Historical slip erosion in catchments under pasture and radiata pine forest, Hawke's Bay hill country. NZ J. For. 40(4):29–33.

Forest and Wood Products Research and Development Corporation. 1996. Evaluation of Santiago Declaration (Montreal Process) Criteria and Indicators for Australian Commercial Forests. FWPRDC Technical Publ 1. For. and Wood Products Res. and Dev. Corp., Gold Coast, Australia.

Johnson, D.W. 1992. Effects of forest management on soil carbon storage. Water Air Soil Pollut. 64:83–120.

Logging Industry Research Organization. 1993. New Zealand Forest Code of Practice. 2nd ed. New Zealand Logging Industry Res. Organ., Rotorua.

Lowe, A.T. 1995. Summary of first five years of Tarawera sustainable forestry trial (FR41). NZ For. Res. Inst. Project Record 4570. NZ For. Res. Inst., Rotorua.

Marden, M. and D. Rowan. 1993. Protective value of vegetation on Tertiary terrain before and during Cyclone Bola, East Coast, North Island, New Zealand. NZ J. For. Sci. 23(3):255–263.

McQueen, D., M.F. Skinner, P. Barker, and A. Lowe. 1994. The effects of compaction and ripping on soil physical properties and radiata seedling growth. *In* Proc. 1994 New Zealand Conf. on Sustainable Land Management, Lincoln, New Zealand. 12–14 Apr. 1994. Lincoln Univ., Canterbury, New Zealand.

National Forest Policy Statement. 1992. A new focus for Australia's forests. Commonwealth of Australia, Canberra.

Neary, D.G., and J.W. Hornbeck. 1994. Impacts of harvesting and associated practices on off-site environmental quality. p. 81–118. *In* W.J. Dyck et al. (ed.) Impacts of forest harvesting on long-term site productivity. Chapman & Hall, London.

Nielsen, J. 1989. Hypertext and hypermedia. Academic Press, New York.

New Zealand Forest Owners Association. 1995. Forestry facts and figures 1995. New Zealand Forest Owners Assoc., Wellington.

New Zealand Ministry of Forestry. 1993. A guide to the Forests Amendment Act 1993. New Zealand Ministry of Forestry, Wellington.

New Zealand Ministry of Forestry. 1994. Statistical release (SR 38): Estimate of roundwood removals from New Zealand forests: Year ended 31 Mar. 1994. New Zealand Ministry of Forestry, Wellington.

New Zealand Ministry of Forestry. 1995a. New Zealand report to the (UN) Commission on Sustainable Development, 1995. New Zealand Ministry of Forestry, Wellington.

New Zealand Ministry of Forestry. 1995b. Statistical release (SR 10): New Zealand planted production forests: Selected statistics for the Year ended 31 Mar. 1994. New Zealand Ministry of Forestry, Wellington.

Powers, R.F., D.H. Alban, G.A. Ruark, and A. Tiarks. 1990. A soils research approach to evaluating management impacts on long-term productivity. FRI (Rotorua, NZ) Bull. 159:127–145.

Proe, M.F., H.M. Rauscher, and J. Yarie. 1994. Computer simulation models and expert systems for predicting productivity decline. p. 151–186. *In* W.J. Dyck et al. (ed.) Impacts of forest harvesting on long-term site productivity. Chapman & Hall, London.

Resource Assessment Commission. 1992. Forest and timber inquiry. Final Rep. Australian Gov. Publ. Serv., Canberra.

Raison, R.J., and P. Hopmans. 1995. Soils. p. 27–42. *In* The development of consistent nationwide baseline environmental standards for native forests. JANIS technical working group on Forest Use and Management, Commonwealth of Australia, Canberra.

Raison, R.J., and P.K. Khanna. 1995. Sustainability of forest soil fertility: Some proposed indicators and monitoring considerations. p. 27. *In* Proc. IUFRO World Congress, Tampere, Finland. August 1995. Int. Union of For. Res. Organizations, Vienna.

Ramakrishna, K., and E.A. Davidson. 1997. Intergovernmental negotiations on criteria and indicators for the management, conservation, and sustainable development of forests: What role for forest soils scientists? p. 1–16. *In* E.A. Davidson et al. (ed.) The contribution of soil science to the development and implementation of criteria and indicators of sustainable forest management. SSSA Spec. Publ. 53. SSSA, Madison, WI.

Smith, C.T. 1994. Is plantation forestry good or bad for soils? NZ For. 39(2):19–22.

Smith, C.T. 1995. Environmental consequences of intensive harvesting. Biomass Bioenergy 9 (1–5):161–179.

Smith, C.T., W.J. Dyck, P.N. Beets, P.D. Hodgkiss, and A.T. Lowe. 1994a. Nutrition and productivity of *Pinus radiata* following harvest disturbance and fertilization of coastal sand dunes. For. Ecol. Manage. 66:5–38.

Smith, C.T., A.T. Lowe, P.N. Beets, and W.J. Dyck. 1994b. Nutrient accumulation in second-rotation *Pinus radiata* after harvest residue management and fertilizer treatment of coastal sand dunes. NZ J. For. Sci. 24 (2/3):362–389.

Technical Working Group on Forest Use and Management. 1995. The development of consistent nationwide baseline environmental standards for native forests. Technical Working Group on Forest Use and Management Rep. July 1995. Joint ANZECC-MCFFA Natl. For. Policy Statement Implementation Subcommittee, Commonwealth of Australia, Canberra.

8 Criteria and Indicators of Forest Soils used for Slash-and-Burn Agriculture and Alternative Land Uses in Indonesia

Meine van Noordwijk

ICRAF-Southeast Asia
Bogor, Indonesia

Kurniatun Hairiah

Brawijaya University
Malang, Indonesia

Paul L. Woomer

Tropical Soil Biology and Fertility Programme
Nairobi, Kenya

Daniel Murdiyarso

Southeast Asian Regional Centre for Tropical Biology–Global Change and Terrestrial Ecosystems Southeast Asian Impacts Centre
Bogor, Indonesia

FUNCTIONS OF FOREST SOILS

What are forest soils? Do they differ from agricultural soils? If so, what about agroforestry soils and the soils that are part of slash-and-burn agricultural systems involving a forest fallow? A substantial part of all land currently used for intensive agriculture, in the temperate zone as well as in the tropics, was once a forest soil and was either used for slash and burn agricultural systems (Steenberg, 1993), or was at least cleared from its forest cover by a slash-and-burn method. A substantial part of current forests has at some point in time been cleared for agriculture, and was allowed to revert to natural vegetation. Subsequent agricultural use of former forest soils is partly based on the inheritance of readily available nutrients and easily decomposable soil organic matter (SOM) pools, but also

The Contribution of Soil Science to the Development of and Implementation of Criteria and Indicators of Sustainable Forest Management. SSSA Special Publication no. 53.

on the soil structure with its network of old tree root channels and on the soil biota in as far as they survived the forest clearing operations (Kooistra & Van Noordwijk, 1996). As long as this inheritance lasts, the agricultural soils will maintain a partial forest soil character. Charcoal formed from roots in the topsoil may stay in the exforest soil for a long time; surface-produced charcoal may wash down slope and be incorporated in sedimentation zones.

Forest soils are not fundamentally different from nonforest soils, but the concerns about long-term changes in forest soils may differ from those under more intensive agricultural management. Changes in soil quality can be considered from at least four perspectives:

1. Effects on the sustainability of current land use practices.
2. Direct effects on other environmental compartments via transfers of nutrients, sediment, water and gases.
3. Indirect effects on land use decisions elsewhere if current land use is nonsustainable.
4. Change in options of conversion to other future land uses.

Category 1 is probably the most obvious one, but Categories 2 to 4 also play a role in the debates on forest and forest-derived soils. Nonsustainability of current land use may be due to effects on the abiotic and/or biotic resource base for future production. It may involve, for example, changes in soil physical properties affecting the water balance, changes in the nutrient balance and soil organic matter content affecting plant nutrition, or changes in essential soil biota.

Examples of Category 2 are the much debated functions of forest soils in watershed protection and regulation of stream flows, and the more recently discovered function of forest soils in oxidizing atmospheric CH_4 and thus offsetting the emissions of (nearby) sources of this greenhouse gas.

An example of Category 3 is the concern about unsustainable forms of slash-and-burn agriculture that may lead to a continued hunger for new forest lands (Brady, 1996). Environmental arguments in favor of promoting alternatives to slash and burn are largely based on the values of the forests before conversion and thus on the deflection value of land use practices (Category 3). Deflection is defined as the area of forest **not** converted due to a certain intensified land use practice—it is based on the inverse of the number of people productively engaged (or employed) and satisfying their livelihood needs per ha of a land use system (averaged across the full production cycle). The difference between this value and that for a long-fallow rotation slash and burn system under local circumstances (say 25 ha per person, for 1 ha cultivated per year with a 50 yr fallow rotation and two persons per family) is the deflection value of the intensified land use practice. The counterpoint to the deflection argument, the likelihood that productive and sustainable alternatives to slash and burn may accelerate rather than slow down forest conversion is now discussed as the Pandora's box issue. Only a combination of intensified, sustainable production systems and adequate protection of the edges of the remaining forests will have the desired effect. A major alternative to slash-and-burn agriculture was developed in the beginning of the 20th century in Sumatra (Indonesia) in the form of jungle rubber systems, or rubber agroforests (Gouyon et al., 1993). Rubber trees were initially planted as an enriched fallow species to increase the value of the forest vegetation regrowing after an upland rice (*Oryza sativa* L.) crop. The result was a sustainable and productive land use system, where the value of the rubber rapidly became the dominant part of the system and the first year's food crops an aspect of secondary importance. The productivity of the system, however, attracted an inflow of people to the area and, in combination with logging operations that provide access and large-scale commercial plantations, led to a situation where little natural forest is left on the lowland peneplains of Sumatra (Tomich & Van Noordwijk, 1996; Van Noordwijk et al., 1995). The overall impact of alternative land use systems is thus based on a combination of the deflection value, the effectiveness of forest protection and the likelihood of migration, which in itself depends on the labor market in other sectors of the economy.

Category 4 is more difficult to assess, but comes into play in discussions about soil pollution: even if it is not detrimental for current land use, or leads to leakage to other environmental compartments, it may reduce future options. On agricultural lands, for example, accumulation of heavy metals such as Cu due to the use of large amounts of pig slurry may make the land unsuitable for grazing by sheep at any point in the near future or for growing vegetable crops. The use of persistent pesticides or pesticides that affect particularly sensitive soil organisms, may reduce the options for future land use conversions. If we restrict the discussion to situations where forest soils will remain under forest cover, we may still consider natural forest, some form of managed logging enterprise or plantation forest. Logging practices may affect the future value of the soil for natural forest in a different way from the value for tree plantations.

CRITERIA AND INDICATORS

The Santiago declaration for conservation and maintenance of soil and water resources (Ramakrishna & Davidson, 1998, this publication) lists seven indicators. For a number of indicators such as erosion, loss of soil organic matter, and soil compaction (a, d, e) it is not directly clear how to evaluate them, but they probably fall under Category 1 and relate to future on-site soil productivity. The other indicators (b, c, f, and g) involve elements of Category 2. None of the indicators involves Categories 3 and 4 (unless soil erosion is seen from the perspective of reducing future conversion options). This may not be surprising, as Category 3 is based on human choices at the landscape level and beyond, and thus is hard to attribute to a particular plot. Category 4 also is more difficult to evaluate, as it concerns options rather than realities.

All the indicators refer to changes from historical values rather than absolute standards. This is understandable from the wide variation in any indicator values in undisturbed forest systems across the globe, but it makes is difficult to evaluate the indicators for any particular site, as reliable historic data are an exception rather than rule. The best we may expect to do is have parameter ranges from historical or presently undisturbed sites as reference and note the gross deviations.

In this contribution to the debate on such indicators we draw on the initial results of the Alternatives to Slash and Burn project in Indonesia for a review of measurable indicators and historic data. Criteria for evaluating the impacts of land use on (former) forest soils (Table 8–1) can be grouped by soil function, focusing on the sustainability of land use practices (Category 1), and on externalities or effects on environmental functions of forest soils (Category 2). The measurables for these various functions, however, show a considerable degree of overlap. Many of them are linked with the maintenance of surface mulch and soil organic matter.

Indicators can be discussed at two levels: easily observable phenomena that can be used in rapid assessments, but which are quick and dirty, and real measurables, for which standardized protocols can be made and interpretation schemes for the values obtained. Qualitative field level indicators may be sufficient for monitoring on-site changes by (forest) farmers or other land users. To them the presence of a surface litter layer and clear forest streams may be enough to evaluate the system they work with. Yet, such simple indicators are not sufficient for legally binding commitments that may lead to law-suits. The latter require rigorous laboratory procedures. Even with such procedures, the interpretation of data may not be unequivocal as absolute reference values are lacking for many of the parameters. For example, a debate on how often landslides occur in natural forest landscapes can cast doubt on any data on sediment loads of rivers after forest conversion.

In the rest of this chapter, we will focus on two proposed indicators of the soil organic matter (SOM) content, as this relates to both on-site and off-site functions of forest soils. The SOM content of exforest soils is of crucial importance to the productivity of subsequent crops (Nye & Greenland, 1960; Palm et al., 1996). We will discuss the C-saturation deficit and C_{org} fractionation as two approaches to make changes in soil organic matter more readily quantifiable. For both of these indicators we will give a scientific rationale, a suggested protocol for assessment and review the data available for interpreting results.

CARBON SATURATION DEFICIT

Rationale

Forest soils may lose a considerable part of their soil organic matter content after changes in management or conversion to other land use. Yet, the variation in soil organic matter content between different sites and soils is so large, that it is not easy to find a proper point of reference, to judge whether specific values are lower than would be expected under undisturbed forest conditions. On the basis of Hassink and Whitmore (1997), we propose to use a dimensionless C saturation deficit , C_{sat}, as the difference between the current C_{org} content and a reference content, $C_{org, ref}$, which is supposed to indicate saturation of the C protection capacity as approximated in an undisturbed forest condition.

$$C_{sat} = (C_{org, ref} - C_{org}) / C_{org, ref} = 1 - (C_{org} / C_{org, ref}) \quad [1]$$

Table 8–1. Criteria and indicators for evaluating the impacts of land use on (previous) forest soils in the Alternatives to Slash-and-Burn project.

Soil functions/criteria	Indicators (field level—qualitative)	Measurables (quantitative)
	Maintain on-site productivity	
Maintaining soil as a matrix of reasonable structure (soil capital)	Erosion: absence of gullies, presence of riparian filter strips and other sedimentation zones, soil cover by surface litter or understory vegetation Compaction: pocket penetrometer Structure: ‘spade test’, root pattern	Net soil loss = internal soil loss—internal sedimentation Percentage soil cover, integrated over the year (or over annual rainfall) Bulk density of topsoil Soil macroporosity and H_2O infiltration rates
Water balance: buffering water between supply as precipitation and demand for transpiration	Soil cover and absence of gullies as indicator or infiltration	Water infiltration versus run-off Soil water retention Effective rooting depth
Nutrient balance: buffering nutrients between supply from inside and outside the system and demands for uptake (soil nutrient capital)	Annual **exports** of P and cations as fraction of total and ‘available’ stock Financial value of net nutrient exports as fraction of potential replacement costs in fertilizer	Changes in stocks of plant available nutrients Changes in mineralization potential or size of organic matter pools C-saturation deficit (see text; Eq. [1]) Limiting-nutrient trials
Maintaining essential soil biota, such as mycorrhizal fungi and *Rhizobium* (soil biological capital)	Sporocarps (mushrooms) for ectomycorrhizal species Signs of ‘ecosystem engineers’ among the soil fauna: earthworms, termites	Spore counts for V.A. mycorrhiza Mycorrhizal infection and nodulation in roots in the field and in ‘trap crops’ in the lab
Tolerable levels of pests and diseases	Weed flora Pest outbreaks, pesticide use	Soil seed bank of weedy species Population dynamics of potential pests and their natural enemies
	Landscape/global level	
Providing regular, high quality water	Stream flow response time to rain storms Turbidity of streams	Stream flow amounts and variability Sediment load of streams
Air filter: mitigating net emission of greenhouse gasses	Aboveground C stocks in biomass and necromass	Absence of agro-chemicals in water Soil C stocks relative to soil C saturation deficit Net emissions of N_2O and CH_4
Biodiversity reservoirs: allowing recolonization of depleted neighboring landscape units, and germplasm collection for ex-situ exploitation	Diversity of aboveground vegetation, based on ‘plant functional attributes’ or PFA diversity (Gillison and Carpenter, 1994)	Diversity of soil biota in selected ‘indicator’ groups

The main question now is how to derive a site-specific value of $C_{org, ref}$. In trying to construct such a point of reference, the differences between undisturbed forest soils from different sites should be linked to easily identifiable site (climate) or soil parameters. Within a given climatic zone, commonly measured soil parameters are preferred that can act as proxy for known processes leading to build-up (or breakdown) of soil organic matter. As soil biological properties are not routinely measured, chemical and physical factors should be sought that are associated with protection of soil organic matter from microbial breakdown. The two main candidates are soil texture and soil pH.

Differences in C_{org} concentrations between clay and sandy soils have long been noted and are usually attributed to various degrees of physical protection of organic matter against microbial attack. The CENTURY model, for example, links the decomposition constant for various soil organic matter pools directly to the clay content of the soil (Parton et al., 1994a). Matus (1994) found, however, no difference in the decomposition rates of ^{14}C-labeled crop residues between clay and sandy soils. Similarly, Hassink (1994, 1997) found that texture is not a good predictor of N mineralization on grassland soils. Apparently, once the larger C protection capacity of clay soils is saturated, the turnover rate of new inputs is similar to that in soils with a lower protection capacity. The C-saturation deficit, as defined above, appears to be a good predictor of net N mineralization of grassland soils, with $C_{org, ref}$ derived by regression analysis of C_{org} on texture for forest soils (of the Netherlands, in the application by Hassink, 1997). The scaling used for C_{org} in C_{sat} suggests that the C_{org} content on degraded soils or in subsoil (with a high C saturation deficit) may rise relatively quickly when organic inputs are applied, while a sustained increases in C_{org} under higher C input regimes (for example caused by elevated atmospheric CO_2) is not likely. The physical protection is not absolute, and especially tillage is an effective way of exposing fresh ped surfaces to microbial attack.

Apart from texture, soil pH may be used in establishing $C_{org, ref}$ for tropical forest soils. A negative relation between soil pH and C_{org} was established for forest soils of Sumatra (Indonesia) in the 1930s by Hardon (1936) for the pH range of 3.5 to 5.5, with a possible upward trend at higher pH. Although no clear mechanism for pH related C transformations is known, and models such as CENTURY do not include pH as a soil factor, the relationship found by Hardon merits further exploration.

Protocol

To establish the reference value $C_{org, ref}$ for a given climatic zone, a large data set is needed, e.g., derived from soil surveys, in which C_{org} is recorded as well as current land use (natural forest and other categories), soil texture, pH, elevation (as proxy for temperature) and preferably soil type (by a locally relevant classification system).

It is desirable that historical data are included in the data set, as long as the soil analysis methods have not essentially been modified. Historical soil data can be used to check any long term trends in the C_{org} of forest soils.

In a multiple regression analysis the various soil and site factors can be tested in their ability to predict (or account for) local variations in C_{org} of forest soils.

It may be necessary to use a logarithmic transformation of C_{org} to obtain the homogenous residuals on which regression analysis is normally based. Data transformation was needed as the C_{org} of soils did not have a symmetric normal distribution reflecting multiple mechanisms and an imperfect soil classification scheme (e.g., shallow layers of peat do not classify a soil as histosol, yet cause a very high C_{org} in the topsoil). Back transformation of the fitted values will lead to wide confidence intervals, again reflecting the variability of the data base used.

Example and Data Interpretation

Van Noordwijk et al. (1997) analyzed soil data for Sumatra obtained in the 1980s in the context of the LREP (Land Resources Evaluation and Planning Project) project. Approximately 2800 profile data were found with complete records of soil type (soil taxonomy), land use, texture, and C_{org}. Five broad soil groups emerged from the analysis with significant between- group differences in C_{org}:

Histosols (peat) covering about 10% of Sumatra, but which may contain >90% of all C stored in Sumatran soils,

Andisols and wetland soils (Aquic groups) both contain about 10% (w/w) of C_{org}. On the Andisols C is intimately bound to clay complexes, while in wetland soils, the C is partially protected from decomposition by anaerobic conditions,

Among the remaining mineral soil types, two groups could be identified: relatively fertile upland soils (mainly Inceptisols) and the Oxisols plus Ultisols, with an average C_{org} content of 3.8 and 3.2%, respectively. The differences between all groups were statistically significant in a *t*-test.

In general, the C_{org} content decreases from primary forest, to secondary forest to areas used for tree crops and a slash-and-burn series, comprising food crops, shrub fallow land, and *Imperata* (alang-alang, cogon, or speargrass) grasslands. On the major upland soils, the difference in C_{org} content between land use types is about 0.5% C. At an average bulk density of 1.25 g cm^{-3}, this represents 10 Mg ha^{-1} for a 15-cm top soil layer. Changes in deeper layers may be expected to be less, and the total change is probably less than twice the change estimated from the top layer only. On the Andisols and the wetland soils, larger differences in C_{org} content are observed between land use types, but the smaller number of observations makes comparisons less certain. Potentially, land use effects on C_{org} may be more pronounced on these soils as management reduces the protection of C_{org} when Andisols are tilled and wetland soils drained.

A comparison can be made (Fig. 8–1) with the analysis made in the 1930s of a large data set obtained by Hardon (1936) from Lampung on the southernmost corner of the island. Lampung was then under transformation from forest to agricultural land, a change that today has been virtually completed. For nearly all land use categories, Hardon's data fell within the more recent data for the Inceptisols and the Oxisols plus Ultisols. There is no indication of increase in soil

C storage under forests in the 50-yr time span during which atmospheric CO_2 concentration increased by 20% in this period, from 0.029 to 0.035%. Hardon's average topsoil content over all land uses (3.53 %) in Lampung is close to the average of 3.46% for these soil groups (i.e., excluding volcanic, wetland, and peat soils) for the whole of Sumatra in the 1980s. The sampling frequency of land use types in this type of survey, however, may not be a true reflection of land use change and its effect on soil C storage.

The data set for the 1980s (Fig. 8–2) confirms the relation between soil pH and C_{org} established in the 1930s by Hardon (1936). The combined data show that the lowest C_{org} content can be expected in the pH range of 5.0 to 6.0. Below a pH of 5.0 reduced biological activity may slow down the breakdown of organic matter. Interestingly, most agricultural research recommends lime applications to the pH range of 5.0 to 6.0; this may stimulate breakdown of organic matter and thus contribute to crop nutrition, but possibly at the costs of maintaining the soil organic matter content. Effects of soil pH on C_{org} might be related to a shift from a dominantly bacterial to dominantly fungal soil foodweb.

In a multiple regression analysis, soil texture, pH, soil type, and land use were investigated (Van Noordwijk et al., 1997). The quantitative factors: pH, clay, and silt, had a slope that differs significantly ($P < 0.001$) from zero. The relative weighing factors for clay and silt are 1.4 and 1.0, respectively. The regression coefficient for elevation ($P < 0.01$) and for slope ($P < 0.05$) also were significantly different from zero. In this regression, the effects of elevation are studied separately from the elevational distribution of soil groups. They indicate a positive effect on C_{org} of lower temperatures. Compared with the average contents per soil type and land use, the C_{org} content will decrease 15% per unit

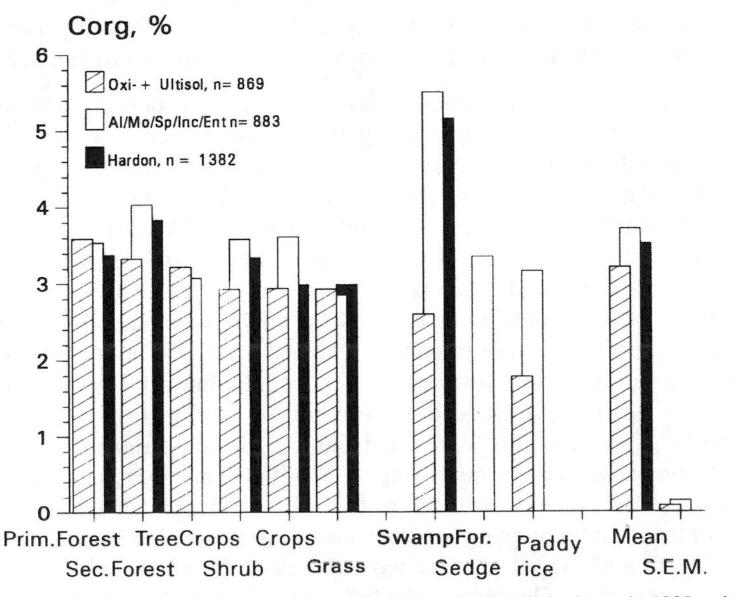

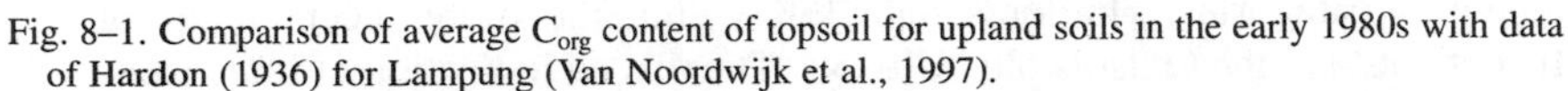
Fig. 8–1. Comparison of average C_{org} content of topsoil for upland soils in the early 1980s with data of Hardon (1936) for Lampung (Van Noordwijk et al., 1997).

increase in pH, increase 1 and 0.7% increase in clay and silt content, respectively, increase by 4% per 100 m increase in elevation and decrease by 0.3% increase in slope. From the regression analysis we propose the following equation for $C_{org,ref}$ in the upper 15 cm of mineral soil from Sumatra under forest cover:

$$C_{org,ref} = \exp(1.256 + 0.00994 \times \text{Clay\%} + 0.00699 \times \text{Silt\%} - 0.156 \times \text{pH} + 0.000427 \times \text{Elevation} + 0.834 \text{ (if soil is Andisol)} + 0.363 \text{ (for swamp forest on wetland soils)}$$

The linearized effect of pH is not very satisfactory in view of the increase of C_{org} above pH 6, but adding a quadratic term for pH to the regression equation did not results in a significant reduction of the residual variance.

As average C-saturation for nonforest land uses, the data set suggests 91% for land with upland crops, 83% for land under tree crops of various types, 85% for *Imperata* grasslands and 81% for young secondary vegetation (shrub). C-saturation decreases with –2.6% per 10% slope. The small reductions for cropped land may come as a surprise. In Sumatra, however, there is little permanently cropped land and most land that is recorded as cropped in a survey was cleared from forest fallows. Also, much of the crop lands is under a no-till system after slash-and-burn land clearing.

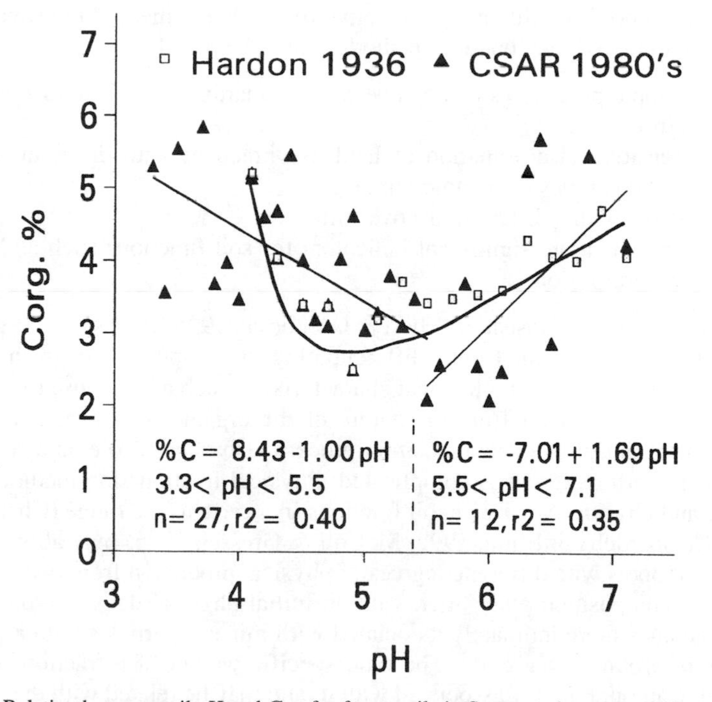

Fig. 8–2. Relation between soil pH and C_{org} for forest soils in Sumatra; the squares refer to data of Hardon (1936), the triangles to recent data (Van Noordwijk et al., 1997).

The concept of a C saturation index may be used in other conditions, but the specific parameters of a regression equation underlying $C_{org,ref}$ are likely to differ, and other parameters may have to be included in the regression equation. The use of models such as the CENTURY model to predict what a site-specific $C_{org,ref}$ would be may be resorted to if data are absent. One should realize, however, that these models may still miss important relationships (such as the effect of soil pH) and may not be universally applicable (Gijsman et al., 1996).

CARBON FRACTIONATION

Rationale

Total soil organic matter content, and its derivatives such as the C saturation index may not be very sensitive indicators, as they change relatively slowly under different management regimes. Measurement of total soil-C is adequate for evaluating C-stocks in the soil, but not for studying soil-C dynamics and functions of soil organic matter over the short term, as only a small part of the total C is responding rapidly to management. Development and testing of improved soil management practices would be easier if more sensitive indicators were available. Models of soil organic matter, such as the CENTURY model (Parton et al., 1994a,b) assume a number of pools of different turnover rates, but the measurement of such pools is still an area of active research and method improvement.

Suitability of fractionation methods should be based on:

1. Simple procedures that can be well standardized and lead to a small lab error,
2. Sensitive discrimination of land use practices with different qualities and quantities of organic inputs,
3. Pools with a different turnover time,
4. Pools that are significant indicators for soil functions such as N and P mineralization.

A large body of research (Ellert & Gregorich, 1995; Blair et al., 1997; Heal et al., 1997) is concerned with differentiating soil C-pools and fresh organic inputs on the basis of their chemical characteristics such as response to mild oxidation agents. But apart from the nature of the organic pools, their nurture or environment during the decomposition process also plays a role, and separating nature and nurture effects is complicated. Physical fractionation methods based on size and physical density of soil fractions in a certain size range (Christensen, 1992; Cambardella & Elliot, 1992; McColl & Gressel, 1995) may allow the distinction of pools with different degrees of physical protection from decomposers. During decomposition plant litter, with an initial physical density around 1.0 g cm^{-3} becomes more intimately associated with mineral particles with a physical density of around 2.5 g cm^{-3}. The total specific weight of a fraction will thus reflect the amount of clay associated with it, and may be related with the turnover time of the fraction. We will here focus on just one of the available methods for physical fractionation and indicate the types of evidence on its value and shortfalls, but it is important to stress that this is an area of ongoing debate and method improvement, Standardization is valuable in that it allows inter-lab comparisons, but it also may be an obstacle to progress. Interpretation of results of any fractionation procedure should be based on local context.

Protocol

A fractionation procedure developed by Meijboom et al. (1995) on the basis of colloidal silica suspensions (Ludox) leads to a light, intermediate, and heavy fraction (Hassink, 1995). This method has been tested in a number of sites in the tropics.

The procedure involves the following steps:

1. Dry sieve the soil through a 2-mm mesh sieve to remove roots and coarse litter particles,
2. Rewet 500 g of soil and leave it for 24 h to equilibrate,
3. Wash the samples on a 150-µm sieve under a gentle stream of water; a 250-µm sieve can be placed on top of the finer one to avoid clogging of the sieve; fine aggregates may be crushed on the coarser sieve during the washing; the silt and clay sized particles passing through the 150-µm sieve are discarded,
4. Collect material from both sieves and separate the coarse mineral sand particles from fractions that contain organic material by decantation in swirling water; this procedure needs further standardization; the mineral fraction is discarded, but may have to be checked on an organic C content while testing the method,
5. The remaining sand-sized fractions are separated into three fractions, by sequential immersion into silica suspensions (Ludox) of two physical densities: 1.13 and 1.3 g cm^{-3}. In the method description by Meijboom et al. a density of 1.37 g cm^{-3}, was used, but its viscosity may cause problems and a suspension of 1.3 g cm^{-3} is preferred (Hairiah et al., 1995); the three fractions are indicated as light (floating on 1.13 g cm^{-3}), intermediate (floating on a 1.3 g cm^{-3} suspension, but not on 1.13 g cm^{-3}), and heavy (not floating on either). The material coming to the surface on a given suspension within a specified time is scooped off, rinsed and dried. Its total C and N content can be measured by conventional means.

For further studies the silt and clay fraction may be collected and yield additional information (Hassink, 1997). Full standardization of the method and comparisons of results between laboratories have not yet been achieved and details of the methods (use of soil dispersion agents prior to washing, mesh size of the sieves used, density used to separate intermediate and heavy fractions) may differ between the various publications. Results on the second, third, and fourth criterion for a fractionation scheme are promising, however.

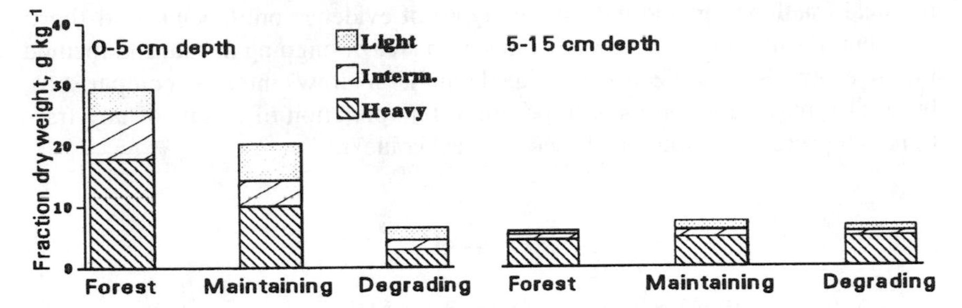

Fig. 8–3. Soil organic matter fractions, based on the Ludox size–density fractionation scheme (Meijboom et al., 1995), for three groups of land use practices in the North Lampung benchmark area of the Alternatives to Slash and Burn project in Indonesia (Hairiah et al., 1995).

Data Interpretation: Indicator Value of Ludox Fractions

In a survey of the organic matter content of topsoil in the N. Lampung benchmark area of the Alternatives to Slash and Burn project in Sumatra (Indonesia; Fig. 8–3), three groups of land use practices could be differentiated on the basis of Ludox fractions of the top 5 cm of mineral soil:

1. Forest (remnants of logged-over primary and various types of secondary forest)
2. SOM-maintaining practices: woodlots, forest plantations established with slash-and-burn land clearing, home gardens, and unburnt *Imperata* grasslands,
3. Degrading lands: burnt *Imperata*, sugar cane plantations with annual burning of residues and forest plantations established with bulldozer land clearing.

For the second category of land use systems the sum of the Ludox fraction (g kg^{-1}) in the top 5 cm of soil may still decrease by about 20 to 30% from the forest level. Under degrading situations, the data suggested that 8 to 10 yr after opening the forest, the sum of the Ludox fraction decreased by 70 to 80%. In the 5- to 15-cm depth layer, however, the converted forest sites exceeded the forest. Total content of the Ludox fractions (in g kg^{-1} of soil) for this second layer is only 20 to 50% of that in the top 5 cm. In the 5- to 15-cm soil layer the heavy fraction is dominant compared with the light and intermediate fraction in dry weight. In as far as the sum of the fractions can be used as indicator, rather than the fractions per se, the procedure could be simplified. Evidence so far is not unequivocal. In a study of long term soil fertility experiments in Kenya (Kapkiyaga, 1996, personal communication), however, results for the simpler particulate organic matter (POM) method (Anderson & Ingram, 1993) based on size only were not improved by a subsequent density fractionation. The POM method, however, uses 50 μm as smallest sieve size and the Ludox method as defined by Meijboom et al. (1995) uses 150 μm.

Table 8–2. Decomposition constants for forest soil organic matter (C_{org}) and for the Light, Intermediate and Heavy fraction of macro-organic matter obtained with the Ludox method based on an analysis of $\Delta^{13}C$ of soil organic matter 1–10 years after conversion of forest to sugarcane in Lampung (Indonesia) (Hairiah et al., 1995).

Fraction	Decomposition constant, k (yr^{-1})	Standard error of estimated k (S.E.)	Percentage of variance accounted for (R^2)
C-L, light	0.194	0.026	91.3
C-L, Intermediate	0.185	0.049	72.6
C-L, Heavy	0.142	0.013	96.1
C-L, Total	0.168	0.024	90.3
Corg	0.082	0.029	57.8

Turnover Time of Ludox Fractions

A direct assessment of the turnover of these Ludox fractions of the original forest soils was obtained from a chronosequence of sites where forest had been converted to sugarcane in the past 10 yr (Hairiah et al., 1995). Analysis of the stable C isotope ratio $^{12}C/^{13}C$ of the Ludox fractions allowed distinction of the organic matter in the three fractions derived from the forest vegetation (a C3 photosynthetic pathway) and from the sugar cane (with a C4 photosynthetic pathway). From these time series decomposition constants could be derived (Table 8–2), for the total C_{org} pool as well as for the various fractions. These decomposition parameters, however, are a net effect of transformations between pools and decomposition (release of CO_2); current data do not allow a full separation of inter-pool conversions. On the basis of the apparent turnover time the light and intermediate fraction can be clearly distinguished from the heavy fraction and the total C_{org} pool, and the regression lines for the decay of the various fractions where better defined (larger R^2 value) than for C_{org}); however, the differences in turnover rate between the fractions are smaller than one might expect. Ten years after forest conversion 25, 40, and 60% of the light, intermediate, and heavy fractions, respectively, still has a forest C signature. These results may be the best indication so far of the gain in information if physical fractionation is taken beyond size as criterion. We do not yet have, however, a satisfactory parametrization for the whole decomposition cascade and inter-pool transformations in the soil (Matus, 1994).

Functional Significance of Ludox Fractions

Barrios et al. (1996, 1997) tested the Ludox method in the analysis of sequential agroforestry systems in Kenya and Zambia and found that the light and intermediate fractions obtained with this method appear to be the most important ones for the N mineralization in the first year after the fallow. When the various Ludox fractions obtained from a range of land use practices in Lampung (Sumatra) with different qualities and quantities of organic inputs were incubated, the light fraction appeared to have the highest specific P mineralization rate (Fig. 8–4).

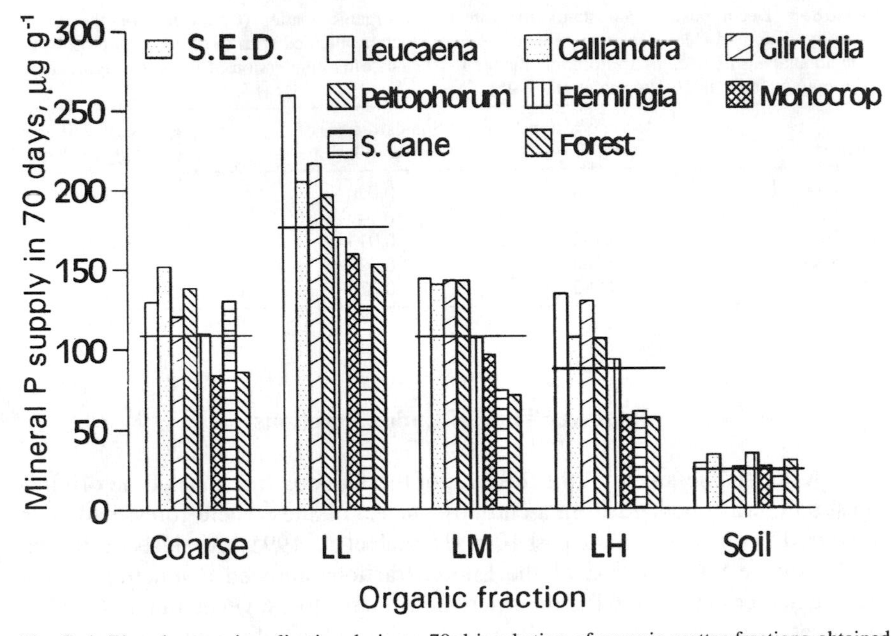

Fig. 8–4. Phosphorus mineralization during a 70-d incubation of organic matter fractions obtained from a range of land use practices with different qualities and quantities of organic inputs in North Lampung.

Conclusion on Ludox Method

The Ludox fractionation method gives more sensitive indicators for studying C dynamics than total soil C_{org}, especially for the 0- to 5-cm depth. Interlaboratory variability in methods, valuable as it may be in the method development stage, makes it difficult to use the method for any legally binding purposes. It is best seen as a research tool as yet. Other approaches, such as the flotation method used by Vanlauwe et al. (1996) should be further compared with the Ludox method.

OTHER INDICATORS OF ON SITE PRODUCTIVITY

For most of the indicators and measurables of soil qualities to be used for on-site productivity [Table 8–1, well-established methods and protocols exist (Anderson & Ingram, 1993; Kooistra & Van Noordwijk, 1996; Hall, 1996]. For rapid assessments, the thickness of the litter layer (rather than presence–absence of surface litter) can be used as indicator, although it should be evaluated with knowledge of the turnover rate and seasonality of litterfall. Spatial variability in litter deposition rates in rain forests may be higher than variability in turnover rates. Burghouts (1993) measured total aboveground fine litter production of 11 Mg ha^{-1} yr^{-1} for a Dipterocarp forest on an acid soil in Sabah. Litter mass on the forest floor was highly variable and varied from a layer of one or two leaves thick to a well-developed layer, 10- to 15-cm thick. He showed that local patches with a thick litter layer were not caused by a slower decomposition rate, but by differences in litter deposition rates. The expected lifetime of surface litter differs between climatic zones and depends on the chemical and physical quality of the litter as well as on soil macrofauna. If surface litter accumulates in plantation forests whereas it decomposed more rapidly in the mixed natural forest the plantation replaced, a further analysis is needed. Perhaps important groups of the soil fauna have not been able to adjust to the change in litter regime.

Assessments of biodiversity for belowground organisms are scarce, especially in the humid tropics. As rapid assessment method for aboveground plant diversity (Gillison & Carpenter, 1994) are available, a first approach is via a presumed relationship between above and belowground diversity in a range of forest-derived land use systems. Measurements of soil biodiversity in the Alternatives to Slash and Burn project include an assessment of surface litter macrofauna, macrofauna in the top 30 cm, general microbial properties of the topsoil, most probable number (MPN) counts of *Rhizobium* and P-solubilizing bacteria, mycorrhizal spore counts and mycorrhizal and *Rhizobium* infection of selected trap crops. Initial results showed large effects of land use on litter fauna, but comparatively small effects on organisms in the mineral soil. Belowground biodiversity indicators apparently respond more slowly to a change in forest cover than their aboveground counterparts. A further refinement of the indicators may be needed, however.

There is remarkably little detailed evidence that agricultural land use in its various degrees of intensity results in a loss of biodiversity in soil (Giller et al., 1997). A rare exception is that of the changes in earthworm populations on conversion of tropical rainforest to pasture in the Amazon Basin, where a single species survives (Fragoso et al., 1997) that leads to soil compaction due to its massive surface casting activity. In this example a reduction in diversity is coupled to and presumably responsible for a loss of function and agricultural productivity. In other cases soils can be mistreated to a remarkable extent and yet crops continue to support crop yields close to their theoretical maximum. Thus the interpretation of soil biological properties as sustainability indicators will require more background data than currently available.

INDICATORS OF LANDSCAPE OR GLOBAL FUNCTIONS OF FOREST SOILS

Functions of forest soils in modulating the water balance downstream are well-known; however, extrapolation between climatic zones of these functions may have lead to an overestimate of these roles in the humid tropics: if rainfall is so high that the whole soil system remains near saturation (with real or perched watertables near the surface), a fairly rapid transmission of rain storms to areas downstream can be expected, either by surface or subsurface flow of water, even under full forest cover. Recent developments in landscape level modeling of runoff and infiltration, indicate a scale dependence of effects of forests on the water balance and indicate that location of forest cover in a watershed is at least as important as the percentage forest cover *per se* (De Ridder et al., 1996).

The air filter function of forest soils and their slash-and-burn derivatives has been mostly discussed from a C storage point of view (Tinker et al., 1996). Opportunities for further C storage are generally limited in topsoils, as the fairly small C saturation deficits indicate. Potentially larger opportunities exist in the subsoil, as those layers may be well below their C-saturation.

Surveys of net CH_4 and N_2O emissions as part of the Alternatives to Slash and Burn project indicated that forest soils can act as a CH_4 sink of considerable strength. According to initial measurements in Sumatra (Van Noordwijk et al., 1995), 1 ha of forest may off-set the CH_4 emissions of about 20 ha of rice paddy. Direct CH_4 exchange, however, may be especially relevant where slash-and-burn land clearing is practiced in a forested landscape mosaic. Smouldering fires release substantial amounts of CH_4, but not all of this will reach the atmosphere if there is enough under-used CH_4 oxidation capacity nearby. The CH_4 oxidation function of forest soils appears to remain intact in forest-like conditions such as found in the rubber agroforests, but it is diminished in more intensive land use types. Further research on this topic may aim to distinguish which soil factor (bulk density as an indicator of macropositiy and hence CH_4 diffusion into soils, specific C fractions or N pools that may modulate the activity of methanotrophe bacteria, or the population size of this group) is responsible for differences between soils of different land use categories.

N_2O emissions appear to be linked to mineral N concentrations. The initial surveys indicated higher emissions from forest soils than for some of the degraded environments in *Imperata* grasslands. Net CH_4 and N_2O emissions can be combined with the changes in the C balance to estimate the net radiative forcing effect of various land uses.

Overall, the trajectory from common sense, simple indicators to protocols that are sufficiently tight to survive courtroom tests and become part of legally binding commitments, is extremely arduous. Whereas standardization of methods is necessary to make cross site and across-time comparisons, development and fine tuning of new methods also is required to keep up with increasing demands from society at large to guarantee that public functions of forest soils will be maintained in the future.

ACKNOWLEDGMENTS

The authors wish to acknowledge the input of several colleagues in the Alternatives to Slash and Burn (ASB) project in developing the list of criteria and indicator; the ASB project is financially supported by the Global Environment Facility (GEF).

REFERENCES

Anderson, J.M., and J.S.I. Ingram (ed.). 1993. Tropical soil biology and fertility. A handbook of methods. CAB Int., Wallingford, England.

Barrios E., R.J. Buresh, and J.I. Sprent. 1996. Nitrogen mineralization in density fractions of soil organic matter from maize and legume cropping systems. Soil Biol. Biochem. 28:185–193.

Barrios E., F. Kwesiga, R.J. Buresh, and J.I. Sprent. 1997. Light fraction soil organic matter and available nitrogen following trees and maize. Soil Sci. Soc. Am. J. 61:826–831.

Blair, G.J., R.D.B. Lefroy, B.P. Singh, and A.R. Till. 1997. Development and use of a carbon management index to monitor changes in soil C pool size and turnover rate. p. 273–281. *In* G. Cadisch and K.E. Giller (ed.) Driven by nature: Plant litter quality and decomposition. CAB Int., Wallingford, England.

Brady, N.C. 1996. Alternatives to slash-and-burn: a global imperative. Agric. Ecosyst. Environ. 58:3–11.

Burghouts, Th.B.A. 1993. Spatial heterogeneity of nutrient cycling in Bornean rain forest. Ph.D. diss. Free Univ. of Amsterdam, the Netherlands.

Cambardella, C.A., and E.T. Elliot, 1992. Particulate organic matter across a grassland cultivation sequence. Soil Sci. Soc. Am. J. 56:777–783.

Christensen, B.T. 1992. Physical fractionation of soil and organic matter in primary particle size and density separates. Adv. Soil Sci. 20:1–90.

De Ridder, N., T.J. Stomph, and L.O. Fresco. 1996. Effects of land use changes on water and nitrogen flows at the scale of West African inland valleys: A conceptual model. p. 367–381. *In* P.S. Teng et al. (ed.) Application of systems approaches at the farm and regional levels. Kluwer, Dordrecht, the Netherlands.

Ellert, B.H., and E.G. Gregorich. 1995. Management-induced changes in the actively cycling fractions of soil organic matter. p. 119–138. *In* J.M. Kelly and W.W. McFee (ed.) Carbon forms and functions in forest soils. SSSA, Madison WI.

Fragoso, C., G.G. Brown, J.C. Parton, E. Blanchart, P. Lavelle, B. Pashanasi, B. Senapati, and T. Kumar. 1997. Agricultural intensification, soil biodiversity and agroecosystem function in the tropics: The role of earthworms. Appl. Soil Ecol. 6:17–35.

Giller, K.E., M.H. Beare, P. Lavelle, A.M.N. Izac, and M.J. Swift. 1997. Agricultural intensification, soil biodiversity and agroecosystem function. Appl. Soil Ecol. 6:3–16.

Gijsman, A.J., A. Oberson, H. Tiessen, and D.K. Friesen. 1996. Limited applicability of the CENTURY model to highly weathered tropical soils. Agron. J. 88:894–903.

Gillison, A.N., and G. Carpenter. 1994. A generic plant functional attribute set and grammar for vegetation description and analysis. CIFOR Working Pap. 3. Centre for Int. For. Res., Bogor, Indonesia.

Gouyon, A., H. de Foresta, and P. Levang. 1993. Does 'jungle rubber' deserve its name? An analysis of rubber agroforestry systems in southeast Sumatra. Agrofor. Syst. 22:181–206.

Hairiah, K., G. Cadisch, M. van Noordwijk, A.R. Latief, G. Mahabharata, and Syekhfani. 1995. Size-density and isotopic fractionation of soil organic matter after forest conversion. p. 70–87. *In* A. Schulte and D. Ruhiyat (ed.) Proc. Balikpapan Conf. on Forest Soils.

Hall, G.S. (ed.). 1996 Methods for the examination of organismal diversity in soils and sediments. CAB Int., Wallingford, England.

Hardon, H.J. 1936. Factoren, die het organische stof-en het stikstof-gehalte van tropische gronden beheerschen (Factors controlling the organic matter and the nitrogen content of tropical soils). Landbouw XI (12):517–540.

Hassink, J. 1994. Effects of soil texture and grassland management on soil organic C and N and rates of C and N mineralization. Soil Biol. Biochem. 26:1221–1231.

Hassink, J. 1995. Density fractions of soil macroorganic matter and microbial biomass as predictors of C and N mineralization. Soil Biol. Biochem. 27:1099–1108.

Hassink, J. 1997. The capacity of soils to preserve organic C and N by their association with clay and sand particles. Plant Soil 191:77–87.

Hassink, J., and A.P. Whitmore. 1997. A model of physical protection of organic matter in soils. Soil Sci. Soc. Am. J. 61:131–139.

Heal, O.W., J.M. Anderson, and M.J. Swift. 1997. Plant litter quality and decomposition: An historical overview. p 3–30. *In* G. Cadisch and K.E Giller (ed.) Driven by nature: Plant litter quality and decomposition. CAB Int., Wallingford, England.

Kooistra, M.J., and M. Van Noordwijk. 1996. Soil architecture and distribution of organic carbon. Adv. Soil Sci.15–57.

Matus, F.J. 1994. Crop residue decomposition, residual soil organic matter and nitrogen mineralization in arable soils with contrasting textures. Ph.D. diss. Wageningen Agricultural Univ., Wageningen, the Netherlands.

McColl, J.G., and N. Gressel. 1995. Forest soil organic matter: Characterization and modern methods of analysis. p. 13–41. *In* J.M. Kelly and W.W. McFee (ed.) Carbon forms and functions in forest soils. SSSA, Madison, WI.

Meijboom, F.W., J. Hassink, and M. van Noordwijk, 1995. Density fractionation of soil macroorganic matter using silica suspensions. Soil Biol. Biochem. 27:1109–1111.

Nye, P.H., and D.J. Greenland. 1960. The soil under shifting cultivation. Commonwealth Bureau of Soils Tech. Comm. 51, Commonwealth Bureau of Soils, Harpenden, England.

Palm, C.A., M.J. Swift, and P.L. Woomer. 1996. Soil biological dynamics in slash-and-burn. Agric. Ecosyst. Environ. 58:61–74.

Parton, W.J., D.S. Ojima, C.V. Cole, and D.S. Schimel. 1994a. A general model for soil organic matter dynamics: Sensitivity to litter chemistry, texture and management. p. 147–167. *In* Quantitative modeling of soil forming processes. SSSA Spec. Publ. 39. SSSA, Madison WI.

Parton, W.J., P.L. Woomer, and A. Martin. 1994b. Modelling soil organic matter dynamics and plant productivity in tropical ecosystems. p. 171–188. *In* P.L. Woomer and M.J. Swift (ed.) The biological management of tropical soil fertility. Wiley, Chichester, England.

Steenberg, A. 1993. Fire-clearance husbandry: traditional techniques throughout the world. Poul Kristensen, Herning, Denmark.

Tinker, P.B., J.S.I. Ingram, and S. Struwe. 1996. Effects of slash-and-burn agriculture and deforestation on climate change. Agric. Ecosyst. Environ. 58:13–22.

Tomich, T.P., and M. van Noordwijk. 1996. What drives deforestation in Sumatra? p. 120–149. *In* Benjaven Rerkasem (ed.) Montane Mainland Southeast Asia in transition. Chiang Mai Univ., Chiang Mai, Thailand.

Vanlauwe, B., M.J. Swift, and R. Merkx. 1996. Soil litter dynamics and N use in a leucaena [*Leucaena leucocephala* Lam (De Wit)] alley cropping system in southwestern Nigeria. Soil Biol. Biochem. 28:739–749.

Van Noordwijk, M., T.P. Tomich, R. Winahyu, D. Murdiyarso, S. Partoharjono, and A.M. Fagi (ed.). 1995. Alternatives to slash-and-burn in Indonesia. Summary Rep. of Phase 1. ASB-Indonesia Rep. 4. ASB-Indonesia, Bogor.

Van Noordwijk, M., P. Woomer, C. Cerri, M. Bernoux, and K. Nugroho. 1997. Soil carbon in the humid forest zone. Geoderma 79:187–225.

9 Epilogue

Eric A. Davidson and Kilaparti Ramakrishna
Woods Hole Research Center
Woods Hole, Massachuesetts

Mary Beth Adams
USDA Forest Service
Parsons, West Virginia

The development of criteria and indicator (C&I) processes occurred rapidly between 1992 and 1997. Implementation has only recently begun, and where these various C&I have been tested, their utility has been limited. Moreover, before these C&I processes can be used as effective international policy instruments, the differences and similarities among them must be carefully studied, and negotiations will be needed to harmonize their various approaches. Before such negotiations take place, however, perhaps a more urgent need is to subject the C&I to rigorous evaluation and refinement by scientists from all of the relevant disciplines, both within and outside of government agencies.

The chapters in this volume sample only a modest fraction of the diversity of soils, environmental conditions, and socio-economic situations in the world's forests to which the proposed C&I processes will be applied. The same can be said for the selection of the forest scientists, with their personal perspectives and professional experiences, represented by this distinguished, but necessarily small group of authors. Despite this tremendous diversity among forest ecosystems and among the ways humans view and use them, at least two common themes emerge from this analysis:

- The development of criteria and indicators is worthy of considerable effort on the part of scientists and policy makers, and the specific set of Montreal Process C&I analyzed in this volume enumerate laudable goals for forest management, conservation, and sustainable development.
- Implementation of the criteria and indicators on national and local levels initially will be crude at best because of lack of historical data suitable as a baseline for comparisons, lack of monitoring programs of sufficient geographic, temporal, and scientific scope, and insufficient experience in relating measurements of soil properties to forest health in a consistent, coherent, scientifically defensible, and tractable manner.

The Contribution of Soil Science to the Development of and Implementation of Criteria and Indicators of Sustainable Forest Management. SSSA Special Publication no. 53.

While the first theme is positive and the second appears negative, the tone that emerges from these chapters is one of enthusiasm for a challenge. Saying that we have insufficient experience to implement C&I on a scientific basis is not to say that we have no experience. On the contrary, this volume documents the considerable scientific basis upon which the C&I approach might be based to achieve the goals of management, conservation, and sustainable development of forests.

The Montreal criteria and indicators are largely *motherhood* statements. No one would deny that soil organic matter is generally good and that erosion and toxic pollutants are generally bad. The devil is in the details. Many of these devilish details regarding potential implementation of the motherhood goals are described in the preceding chapters. In addition to identifying problems, several constructive suggestions are offered that should be considered as C&I are refined.

Perhaps most importantly for forest soil scientists, the criteria and indicators process is welcomed by these representative authors because it may help provide the impetus for giving forest soils the attention that they deserve, but that has hitherto been largely withheld. As eloquently explained by these scholars, soils play critical roles in forest ecosystems, and measuring changes in soil properties has the potential to tell us a great deal about long-term sustainability of forest management practices and uses. The soils, themselves, should be managed as a sustainable resource. If implementation of criteria and indicators places an emphasis on studying and monitoring forest soils, however crudely at first, the long-term health of the world's forests may eventually benefit.

One of the take-home messages of this collection should be that an international agreement to abide by some sort of harmonized list of criteria and indicators will not, by itself, ensure enlightened management of forests. Success will come only if the agreement is followed by substantial commitment on the parts of national and local governments, institutions, scientists, and managers to work out the crucial details of how lists of criteria and indicators can be made to work as a positive influence on forest management.